MATRIX OF CREATION

MATRIX OF CREATION

Sacred Geometry in the Realm of the Planets

RICHARD HEATH

Inner Traditions
Rochester, Vermont

Inner Traditions
One Park Street
Rochester, Vermont 05767
www.InnerTraditions.com

Originally published in the United Kingdom in 2002 by Bluestone Press under the title *The Matrix of Creation: Technology of the Gods*

LIBRARY OF CONGRESS CATALOGING-IN-PUBLICATION DATA

Heath, Richard, 1952–
The matrix of creation : sacred geometry in the realm of the planets / Richard Heath.
p. cm.
Includes bibliographical references and index.
ISBN 0-89281-194-3 (pbk.)
1. Cosmology. I. Title.
BD511.H43 2004
113—dc22

2004006974

Printed and bound in the United States at Capital City Press

10 9 8 7 6 5 4 3 2 1

Text design and layout by Priscilla Baker
This book was typeset in Sabon and Avenir with Bauhaus and Galaxy as the display typefaces

ACKNOWLEDGMENTS

The main ideas in this book came through meditations employing a Casio one-memory scientific calculator. My brother Robin set the ball rolling for this type of astronomical research and was a frequent collaborator in a field of two. His experiences in book production proved invaluable in publishing the first edition. For this I am most grateful.

Without the matrix of my own creation, this book would not exist. Therefore, sincere thanks go to my father, the late Professor Fred Heath, who would have had fun with this material, and to my ever-supportive mother, Barbara.

Additional perspectives on these ideas were derived from the late J. G. Bennett's work. Special thanks go to Robert Gibson and Tony Blake. The writers of classic works in related fields who influenced the book are Robert Graves, George Gurdjieff, Ernest McClain, John Michell, Peter Ouspensky, Giorgio de Santillana, R. A. Schwaller de Lubicz, Alexander Thom, and John Anthony West; more recently, Tony Blake, Robin Heath, John Neal, and William Sullivan. John Martineau influenced this book through his experimentation in presenting such material.

Finally, complete thanks to my wife, Jane, who both terrorized and tolerated this consolidation of the male mysteries.

The Bush Barrow Lozenge. This gold artifact was discovered in 1808, near Stonehenge. Measuring seven inches tall, it lay across the chest of the skeleton of a tall man who may have been an astronomer/priest. Its ninefold geometry produces acute angles that define the range of sunrises at the latitude of Stonehenge. The obtuse angles match the extreme risings and settings of the Moon. (Illustration courtesy of Dr. Euan MacKie from the artifact in Devizes Museum, Wiltshire, England.)

CONTENTS

Our Moon

FOREWORD

There are many experiences that unite humankind, but none so fundamental than sharing the same planet. Earth hangs in space like a living jewel, glittering through space and time, and although we tiny humans may squabble about many things, we must all surely agree that the lunar month takes about thirty days to complete, that the seasons are complete in a year of 365 days, and that we all take a single rotational journey on Earth during a single day. Each of these cycles is connected with a certain number of days. These numbers present themselves as unchanging to any observant soul.

We also share the same moon, and from the very word *moon* derive words such as *metrology* and *mensuration.* The Moon's phases initiated our ancestors into counting and measuring time using days, months, and years. This culminated in their ability to predict eclipses.

In our modern era, since Kepler and Newton, we have come to understand that the forces that maintain and sustain Earth in its motion are defined by simple mathematical relationships. Through this understanding we are now able to navigate spacecraft around the other jewel-like planets that share our solar system, and accurately measure the distances, speeds, forces, and masses that govern and regulate the celestial rhythms to which we all dance. But modern science, having discovered mechanisms underpinning planetary motion, paradoxically dismisses any idea that the solar system is other than a random assemblage of planets orbiting the Sun. Yet, if simple numerical formulae can describe the lawful behavior of planets within the solar system, then surely wisdom demands that we inquire further about the nature of the divine lawmaker.

Ancient civilizations held sacred beliefs that supported their prime quest to better understand the divine lawmaker. In examining these civilizations, one will discover what are today termed the traditional arts—astronomy; its sister, astrology; sacred geometry; number sciences; and musical harmony. From such traditions come the sayings "As above, so below," "All is number," and "God is a geometer." But we no longer really believe these adages; we do not walk the ancient talk anymore.

Matrix of Creation is an exciting exploration of the numerical relationships hidden within the solar system. Richard Heath shows that there is a sublime harmony pervading the motions of the Moon and planets. He has pioneered a new science and has shown, using the axioms of science, that it is possible to find the qualities of the divine lawmaker via the quantities expressed in the skies. The beauty, cohesion, and humor of the numerical relationships he reveals are found to be firmly rooted within the fertile soil of the traditional arts and will be familiar to anyone who has undertaken even a perfunctory study of these things. Now they have a new and powerful unfolding.

Richard and I have collaborated for over nine years on the various topics that appear within our books. As brothers, we have shared a great deal more than the delights of discovery, and it has been a pleasure to witness Richard evolving his earlier love of astronomy into this book. A new yet strangely familiar landscape has unfolded right in front of our eyes.

The integration of Earth's mysteries—geodesy, geomantics, and metrology—with the core content of myths and legends means that *Matrix of Creation* will appeal to psychologists, mythologists, astrologers, and alternative thinkers in addition to being essential reading for astronomers, mathematicians, metrologists, and historians.

Throughout this book, the reader will discover a complete vindication of the ancient maxim that the study of number, ratio, and proportion is an essential prerequisite of true cosmology. Sir Arthur Eddington, a great cosmologist of the twentieth century, once said, "If you haven't got the numbers then you haven't got the science." Well, here are the numbers and here is an emerging new science, a re-flowering of perhaps the oldest known science.

ROBIN HEATH,
AUTHOR OF *SUN, MOON & EARTH* AND *STONEHENGE*

A model by Johannes Kepler (1571–1630) demonstrating a planetary system based upon simple numbers and geometry. Kepler struggled for many years before discovering the laws of planetary motion, laws that enabled Newton to discover the laws of gravity and momentum, thereby beginning the technological age. Here we see his illustration of the geometry of the orbits of Jupiter and Saturn, the two outermost visible planets in the solar system.

INTRODUCTION

This book interprets our planetary system purely through numbers. These numbers are surprisingly simple and accessible to anyone holding a basic arithmetic education. I will show that the solar system evolved in an orderly, numerical way and cannot be thought of as an accident of nature. This provides substantial insight into the realm of mythology, religious thought, and the traditional arts.

It is no coincidence that this work is poised between the realm of mathematics and the world before numeracy. In the ancient world, numbers had godlike powers of creation through stable numerical relationships. These were the core of ancient science, which developed through simple astronomical observations and by counting events and regular movements of objects in the sky. This science was articulated as mythological stories, calendars, sacred geometry, musical theory, and monumental architecture such as the Great Pyramid and Stonehenge.

The philosophy behind the design of ancient monuments has remained obscure to the modern world because our scientists have become too sophisticated to see the simple celestial relationships in our planetary system. This design implies that a divine being has manifested creation. The primary criticism of prescientific thought is that its gods are merely the emotional or superstitious projection of the primitive mind onto a misunderstood universe.

These criticisms dissolve when these gods are understood as codifications of the design rules of the universe.

Religious faiths maintain belief in a divine creation on the basis of faith rather than proof. Modern science from the seventeenth century onward holds to a model of the world based on material forces that can be proved to exist. This is called materialism. If ancient man proved that the creation is of divine origin by using a numerical model, then prehistoric thought was not based on faith or superstition but was a nonmaterialistic science that included the divine. But such proof has been lost, leaving the monuments of ancient cultures as grandiose fantasies with no apparent application or relevance to the modern world.

This book equips the reader to enter the simpler view held by our ancestors and to reevaluate humanity's role in an evolutionary system based upon numbers. The word *matrix* means an "environment in which something has its origin, takes form, or is enclosed." In mathematics, a matrix is an array of mathematical elements arranged to show the numerical relations among its components. I have appropriated this word to describe the numerically creative context for the universe.

The actual door into the numerical matrix of creation opened for me in September 2000, toward the end of the Jupiter–Saturn conjunction that took place in the summer of that year. This marked the end of nearly nine years during which my brother Robin and I developed tools and collected material that allowed for rapid progress to the conclusions presented here.

This book is the factual basis for a spiritual cosmology and a timeless phenomenon that I call the matrix of creation. By recovering the ancient revelation that the world we inhabit was created by numerical relations described in our planetary system, it comes to the aid of the beleaguered belief that God made the world in seven days. Thus, God becomes celestial motion conveying spiritual virtue through numerical proportions expressed in physical forms.

The clarification of ancient systems of thought is an important part of this story. These systems lack written records either because of widespread destruction or because they were transmitted by oral tradition.

This has given archaeologists free rein to develop a set of speculations based on the physical remains of cultures.

They hypothesize about unknown ancient rituals because these are seen as superstitious and prescientific. However, the legacy of the monuments, so conspicuous in ancient high cultures, is full of numerical and astronomical knowledge that our scientific elite has ignored. Ironically, they search for planets light-years away that may support intelligent life while they overlook the high intelligence of their own ancient colleagues. To continue the work of the ancients requires the rediscovery of archaic numerical procedures.

The Greeks inherited the science of numbers when they interacted with the ancient world during the empire of Alexander the Great (336–323 BCE). Units of measure such as the Imperial yard, foot, and inch are ancient. The ubiquitous seven-day week came from Babylon, and calendars are a venerable form of human knowledge related to the generation of time through the planetary world. It is the planets and their continuing interaction in the sky that are the origin of ancient numeracy from counting to concepts of time.

Though planets generate numbers, they do so differently from the collections of objects on Earth. I can add a brick to a line of bricks and there is simply one more brick. However, the planets create an intensive system of measurement by dividing the sky and each other's motion. The division of a whole forms such measurements. The sky is that fundamental whole, the origin of all numerical and calendrical events. The subtle mathematics of the celestial realm achieves a great deal by simple means. This simple mathematics is the foundation on which modern science was constructed.

We are told repeatedly, and from many ancient cultures, that the ancient world suffered a sustained deluge that left only its barest foundations. Only rumors describe what stood there and how the inhabitants practiced their culture. These ancient practices can be recovered by attuning our minds to the extraordinary.

Our Earth has life, but what is life? It is more than organized matter. It is the highest field of structured information in the universe. One must ask, What is it about life that causes it to arise, have thoughts, create written language, and generate more information? The planetary matrix is the smoking gun in this mystery. Life is possible on Earth  because of the Sun, the Moon, and the other planets. This matrix is the Holy Grail of ancient science, which, through observation of the skies, revealed the order that set the human mind in motion along its recent evolutionary path.

Figure 1.1. The Seven Visible Planets

From top left, Saturn, Jupiter (and its two moons, Io and Europa), Mars, the Moon, Venus, and Mercury. The Sun, the seventh planet in the ancient cosmology, is indicated by the illumination of the rest. This stylized illustration, which is not to scale, shows a small section of Earth in the lower part of the sky.

ONE

CUTTING UP THE SKY

Human life involves mathematical problem solving. For example, if I cut a plank into equal sections to make bookshelves, how long will the bookshelves be?

A plank of eight feet can be divided into three equal parts of two feet plus a fraction of a foot in each piece. The numbers two and three become interrelated as I solve this problem, for the fraction that remains is two thirds of a foot for each piece. Eight feet divided by three yields three pieces of planking two and two-thirds feet in length. That is, each shelf is two feet, eight inches long.

To solve such problems, our brains develop the skills of proportion and rationality before puberty. They combine spatial and intuitive skills with linguistic and analytical ones. Our ancestors possessed these skills for tens of thousands of years.

But the sky is not linear like a plank. It is composed of whole circles such as the path of the Sun during the course of a year, of the Moon during a sidereal month, and of a planet in a synodic year. The rotation of Earth creates another circle. In fact, each day the whole starry firmament rotates once in a clockwise direction above our heads.

Whole circles surround us as if we are the center of all cosmic action. Like the ouroboros snake, they swallow their own tail, for when one celestial cycle is completed, it starts all over again (figure 1.2).

Picture an ancient man or woman witnessing astronomical changes in the sky. How might he or she have comprehended this spectacle? The observer would have noticed that the rotation of Earth is an excellent timepiece. It creates day, night, and the motion of the Sun and stars. This rotation is extremely stable.

This ancient person saw that the progress of the Sun could be measured in days, months, and years. The stars were seen as a fixed pattern over which the motion of the Sun, the Moon, and planets superimpose themselves. The observer measured the repetition of similar events in days. Also, the detailed positions of celestial objects were measured relative to the stars, without which the sky would be cosmically rootless. The measurement of the motions of the Sun and Moon and the cutting of the continuum of the stars by the cycles of the planets came to be known as *time*.

Figure 1.2. The Ouroboros Archetype of Celestial Cycles (Drawn by Robin Heath)

THE DIVIDED SKY

Nature's cycles do not divide into each other exactly. Like our bookshelves, they leave us with fractions of a unit. However, the true whole can be inferred from the part, just as a shelf of two feet, eight inches belongs to an eight-foot-long plank divided by three. In a solar year, there is a fraction more than twelve lunar months. The fraction is just

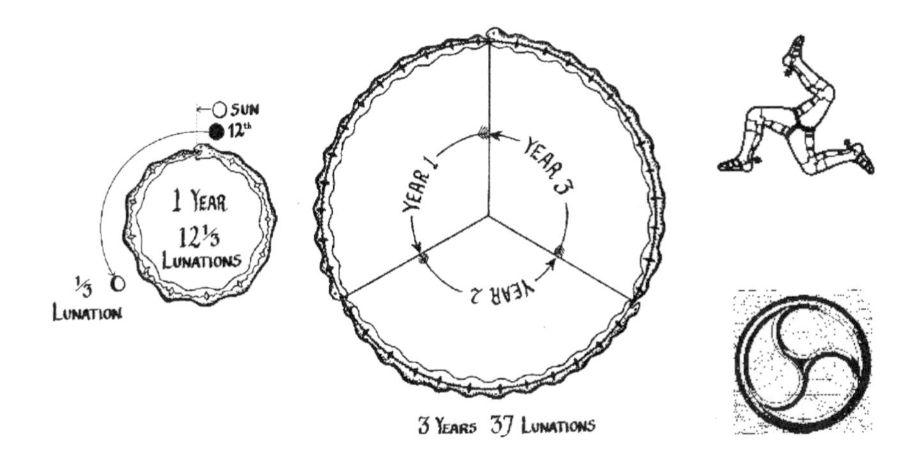

Figure 1.3. The threefold quality inherent in lunar motion as seen in symbolic art. The Manx national insignia (top right) and a typical Celtic church window (lower right) express the geometry of lunar motion. (Drawn by Robin Heath)

over one third of a lunation. So, we say, "Aha! The whole cycle is three solar years in length. From it we can cut three sections, each composed of twelve and one-third lunations—three, one-year sections—with almost no waste." Three years thus contain thirty-seven lunations. After three years, the fractional parts of this cycle add up and thereby disappear (figure 1.3).

The three-year Moon cycle was fundamental to ancient man, since thirty-seven lunar months fit nicely into three years. This created a strong basis for triplicity in the ancient world. It is the first calendrical repeat of the Sun and Moon, accurate to 3.09 days in three years. Better repeats occur after eight years (ninety-nine lunations, accurate to 1.5 days) and nineteen years, the Metonic cycle of 235 lunations (accurate to two hours). All three repeat cycles were used to regulate time in various ancient calendars.

The whole cycle that we know best is the year, because the main seasonal cycle fits within it. In practice, a year is experienced as 365 whole days. I shall call this a practical year. The Sun's annual cycle, the tropical year, which causes the manifestation of the seasons, is a fourfold structure composed of two solstices and two equinoxes. The four seasons are a cultural recognition of this structure as reflected in our use of the word *quarter* in financial analyses. The midpoint of each quarter identifies the quarter days, which are still celebrated as festivals throughout the Celtic lands—Imbolc (February 1), Beltane (May 1), Lughnasa or Lammas (August 1), and Samhain (November 1). The Christian Church later adopted the quarter days as important holy days.

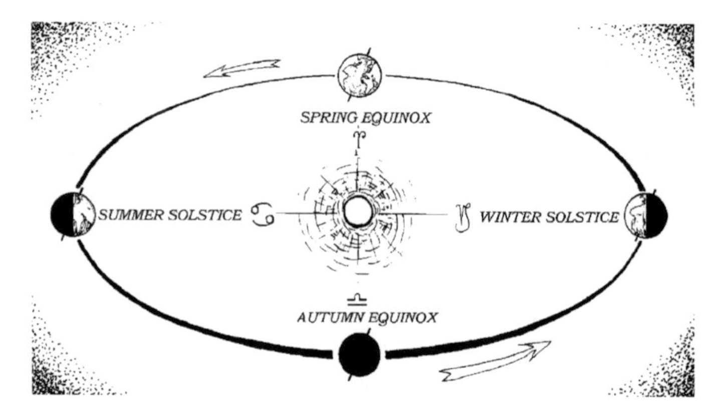

Figure 1.4. Fourfold Division of the Yearly Cycle (Drawn by Robin Heath)

The observable numerical manifestations of the Sun and the Moon can be extended to the planets. Prehistoric cultures chose to represent their knowledge within a spoken tradition through myths, and these almost always included a definite character such as Saturn, Satan, or Set. He is shown with a sickle or harvesting tool and often cuts up a primordial god. He is also called Father Time. This indicates that time is created in some way through his agency. The mechanism of this creation is probably as follows.

The slow yet inexorable motion of Saturn through the night sky is observed quite clearly. About once a year, when Saturn is opposite the Sun, the planet describes a retrograde loop among the stars. Initially the planet becomes stationary with respect to the stars. Then, for a few months, Saturn tracks backward, only to stop again and resume its

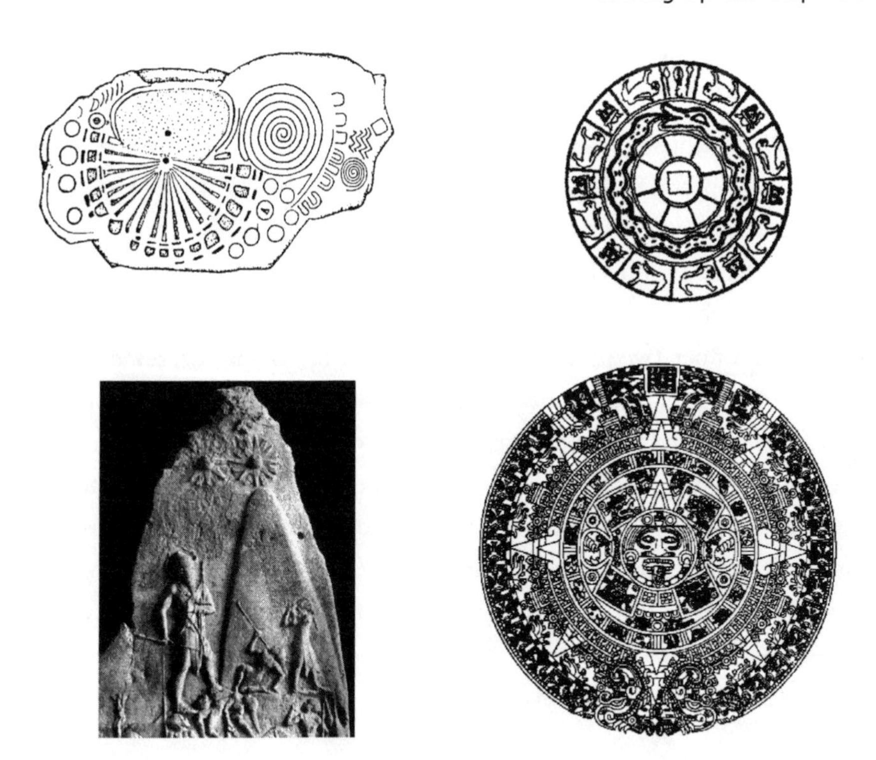

Figure 1.5. Solar Imagery in Ancient and Prehistoric Art
The Sun is depicted throughout ancient cultures as four-, eight-, or sixteen-fold. The curious stone pictogram (top left) is from Knowth, Ireland. The sixteen radial lines ending in squares can be interpreted to match the sixteen-month Stone Age calendar discovered by Professor Alexander Thom. The Stele of Nabus (lower left) depicts a sixteen-fold solar disk over a megalith, dating a victorious battle around 2500 BCE. *Other solar symbols include the swastika or year-snake (top right), various winged wheels, and chariots ridden by heroic sun-gods. The Aztec Sun-wheel (lower right) further embellishes the "four-gated city" of the year.*

forward motion. The time between these loops is called the synodic period of Saturn—378 days.

Saturn's retrograde loops take more than a year to recur but return on the same day of the year every twenty-nine practical years. In that time, Saturn performs almost exactly twenty-eight retrograde loops. The time between loops is therefore twenty-nine practical years divided by twenty-eight, making the synodic period of Saturn twenty-nine twenty-eighths of a practical year. Therefore, a practical year is twenty-eight twenty-ninths (28/29) of Saturn's synodic period of 378 days.

This relationship between Saturn and the practical year is readily observed in practice. It implies the existence of Father Time. His character emerges from direct, measurable experience. The relationship between Saturn and the practical year can be seen in the following equations: 28 x 378 days = 10,584 days; 29 x 365 days = 10,585 days.

The number twenty-eight also is seen in the outer Aubrey circle of Stonehenge as its double, fifty-six, since the Aubrey circle is composed of fifty-six holes. This makes it possible to map Saturn's motion onto Stonehenge. If the whole circle of Stonehenge represents the practical year, a marker representing Saturn would move counterclockwise by two holes every year. Every one and one-twenty-eighth practical years a sun marker would again align with it. After twenty-nine solar years, Saturn as seen from Earth would have moved around the circle only once. This is called a Saturn return.

RESONANCE AND IDENTITY

We must consider two very relevant concepts concerning archaic astronomy.

The first is resonance. The repetitions of astronomical cycles have numeric factors that are felt by life on Earth through climate, light, heat, tides, and bodily cycles. The rate of growth, breeding time, and migratory patterns of animals and the synchronicity of life to climate show how Earth has evolved a harmony with the heavens. Change a cyclical rhythm and evolution will match it in resonance. The ancients appear to have extended this to a paradigm: "All that is in the sky gives rise to that which arises on Earth." Resonant systems are in rhythmic interaction based upon numbers. In other words, numbers resonate throughout the biosphere via the environment to which life adapts. This awakens the human mind to time and numeracy.

The emergence of whole numbers such as twenty-eight and twenty-

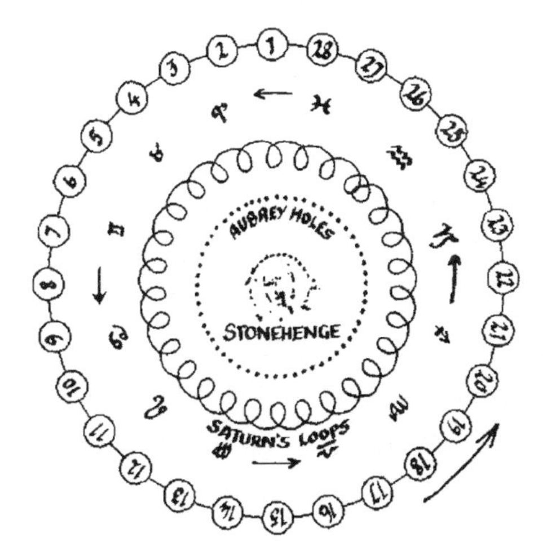

Figure 1.6. Circles of Time

Saturn's twenty-eight loops around the zodiac, which take twenty-nine years to complete, are spaced apart by the same angular distance as the Moon's average motion through the stars in a single day—thirteen degrees. The fifty-six Aubrey holes, forming the earliest constructional phase at Stonehenge, circa 3100 BCE, *provide an analog method of recording the motion of both bodies. An accurate calendar and the prediction of eclipses can readily be undertaken from such a construction. (Drawn by Robin Heath)*

nine that interconnect two major observable cycles gives numbers their own identities. For example, twenty-eight equals four times seven (4 x 7) and fifty-six is eight times seven (8 x 7). On Earth, we have a seven-day week. This identification of the number seven harmonizes heaven and Earth. Calendars usually harmonize the seasons. They also can focus on lunar months or attempt to harmonize the solar and lunar cycles. Ancient calendars often included Venus, Jupiter in passing, and the star Sirius. The archetypal identity of these objects was rooted in a continuum that was essentially numerical and celestial. The challenge of comprehending these resonant identities numerically within time lay at the heart of ancient conceptions of the universe.

THE ROLE OF IDENTITY IN THE CREATION

Cosmos is a word used for both a perfectly harmonious creation and the principle of perfect order, thought to lie behind such harmony. The opposite of perfect order is chaos or perfect disorder. This is experienced in the lowest form of energy, heat, but is seen as beyond or prior to order. For order to manifest as harmony, an entity must transform and organize the dispersed energies of chaos. There is a triad of factors: the extremes of cosmos and chaos mediated by entities that exist in the real world. Any living thing in the biosphere makes order of the chaos within its environment. In doing so, it re-creates itself as a sustainable and recurrent species of life.

The highest scale of identity known to us is our galaxy. At its heart lies a massive black hole creating an ordering field of gravity. Within this, primordial matter is sufficiently structured to facilitate the creation of star-forming clouds of gas and dust. These collapse when swept up in the radiation of massive stars. Gravitational collapse heats this material to temperatures sufficient to convert hydrogen to helium, and a new star is born to burn for billions of years. Around such a star, there may be a solar nebula. Over time, this differentiates into different types of planets due to resonance within the rotating whole. Chaotic collisions are turned into orbiting bodies whose orbits can be stable for billions of years.

Such a planetary system is made up of identities that are objects in space and relationships in time and number. They emerged out of numerical resonance within the early solar system. Their numerical relationships are their foundation of identity in the planetary world. This identity harmonizes cosmos and chaos at a given scale in the cosmic order.

The archetype of the planetary world is that the one is many objects and the many have one relationship.

EARTH AND HER BIOSPHERE

The identity of Earth has two important characteristics. First, when looking up, we see the movement of the other planets as the combined effect of their motion around the Sun and the motion of Earth around the Sun. Only the Sun and Moon follow a continuous onward path—a counter-clockwise motion against the stars. The planets, as the Greek etymology of the word would suggest, "wander" through the sky, sometimes appearing to go backward in the heavens. This retrograde motion is a result of our earthly vantage point.

The second characteristic of Earth's identity is its highly structured biosphere. This new type of identity supersedes the planetary one. The planetary motions seen from Earth suggest that numbers are transformed into forms in time and space. These biological forms are created within a special environment and are infused with numerical resonances created during the formation of our solar system.

This concept of number leading to life was common in ancient thought. It formed a theory of cosmogenesis among the world's mythological traditions. The great city-building cultures built monuments that connected astronomy and numerology and employed the unique proportion called the Golden Mean. This proportion is also found in biological forms.

In the conventional scientific view of things, it is unnecessary that there be any great degree of order in the solar system, especially when seen from the moving platform of Earth. Therefore, the rediscovery of the planetary matrix indicates that a change of paradigm is possible. This paradigm explains how the planetary system came into existence during the formation of the universe.

HOW THE PLANETS CAME TO BE

The Sun is a classic star condensed from the perfect amount of galactic gas and dust. Its birth was triggered by the death and explosion of one or more stars of greater weight.

The Milky Way galaxy is a typical star-making condensation of the original matter of the universe with a massive black hole at its center. Thus, the Galaxy has a dark heart while its billions of solar systems have light ones—the stars around which planets may orbit.

Billions of stars orbit the center of our galaxy but very little structure can be found in their distribution. Stars are the loneliest entities in the universe. Separated by vast distances relative to their size, they rarely approach or influence each other. The separation between solar systems allows the planets of those systems to be influenced only by their sister planets and occasional comet impacts.

Initially, the matter circling a star condenses, through collisions, into planets. After formation, planets move in a friction-free environment like perpetual motion machines. This is very different from our experience of motion on Earth, where movement meets resistance through friction—i.e., collisions—with air, water, or solid surfaces. The orbit of a planet is a constant. This is a planet's defining characteristic. Out of the chaos of cosmic particles, the collapse of the first galaxies, and the formation of stars within galaxies, a dynamic system arose that expresses invariable numerical information for billions of years.

PLANETARY UNFOLDING

The solar system is a finely graded system of light and darkness, heat and cold, and dense and light materials. In time, the nebula that formed our solar system flattened out into a spinning disk. The material destined to become the Sun collapsed until thermonuclear fusion began to generate heat and light by transforming hydrogen into helium. Meanwhile, four solid planets of similar construction coalesced around metal cores. Mercury ended up being mainly a core, Venus and Earth grew to similar mass, and Mars remained rather small. At that distance from the Sun, something else was emerging.

The second phase of planet building generated giants out of gaseous materials in the cooler, outer solar system. Hydrogen and helium gases collapsed into a planet that was nearly a star—Jupiter. It has a sixth the mass needed to create thermonuclear fusion. At twice Jupiter's distance from the Sun, Saturn formed out of hydrogen and helium. This was the visible limit of the solar system to ancient man and it is the effective limit of the planetary matrix.

The orbit of each emerging planet gravitationally affected the other planets as they moved past each other according to a pattern more regular than clockwork. All were heavily affected by massive Jupiter, which has an additional role of "throwing" planetoids and comets from the outer solar system into the realm of the four solid planets.

CREATION OF EARTH'S MOON

Impacts from asteroids and comets, hurled by Jupiter or Zeus, created the tilting of the planets and, consequently, the seasons on Earth. Our moon was generated by a very severe impact that scooped up part of Earth's crust and placed it into orbit. There it absorbed energy from the tides on Earth and settled into an orbital distance that, strangely enough, allowed it to exactly eclipse the Sun.

Meanwhile, the orbits of the planets created numerical interactions with each other as seen from the transformative view of Earth's orbit. The numbers expressed by planetary orbits are transformed into new numbers on Earth. Our planet, the intended common denominator in this matrix, then brought forth life.

A cosmology gleaned from central Asia by G. I. Gurdjieff tells that the galaxies were formed first, the stars second, and then the planets. This left Earth's biosphere as the fourth item of creation. It reflected in its organization all that had gone before. Gurdjieff asserted that a sacred individual accidentally created the Moon during the formation of the planetary system. This points to the action of Jupiter, since its gravity is capable of slinging comets into the inner solar system. According to Gurdjieff's sources, a planet has to evolve life on its surface in order to support a substantial moon. But the truth may be the other way around—a substantial moon is a prerequisite for evolved life.

The numerical revelation in the planetary system, the creation and placement of the Moon, and especially the creation of life can be seen as an act of demiurgic intelligence. In fact, Jupiter was known as the great demiurge in ancient myth. The word *accident* can be defined, positively speaking, as a coincidence that appears accidental. Such a definition

Figure 1.7. The rhythms of the heavens lie at the heart of all ancient sciences. This thirteenth-century illustration depicts the struggle to discover what lies behind the cosmos.

explains why a new level of order entered the physical world from an invisible dimension.

THE PLANETARY MATRIX

The planetary matrix is the set of numerical planetary relationships, seen from Earth, that may have caused the creation of the biosphere.

The numbers of the matrix, which appear as time, fit with traditional forms of wisdom in which the creation sprang from the mind of God. This is the realm of the pure interrelationships of mathematics. Biological evolution resulted in beings with brains that can discover mathematical relationships. On Earth, where numbers are generated by the gods themselves through the recurrence of events such as seasons, months, the appearance of morning stars, and conjunctions of all sorts, the brain is naturally susceptible to counting and numbers.

Jupiter, the creator-god known variously as Brahma, Jehovah, Brihaspati, or Zeus, created a matrix of numeric relationships through the Moon. On Earth, the matrix is seen in a useful calendar of objective time. The relation of events to the location of the planets in the matrix can be seen through such a calendar. This is appropriate for thinking beings who have evolved in the matrix of creation.

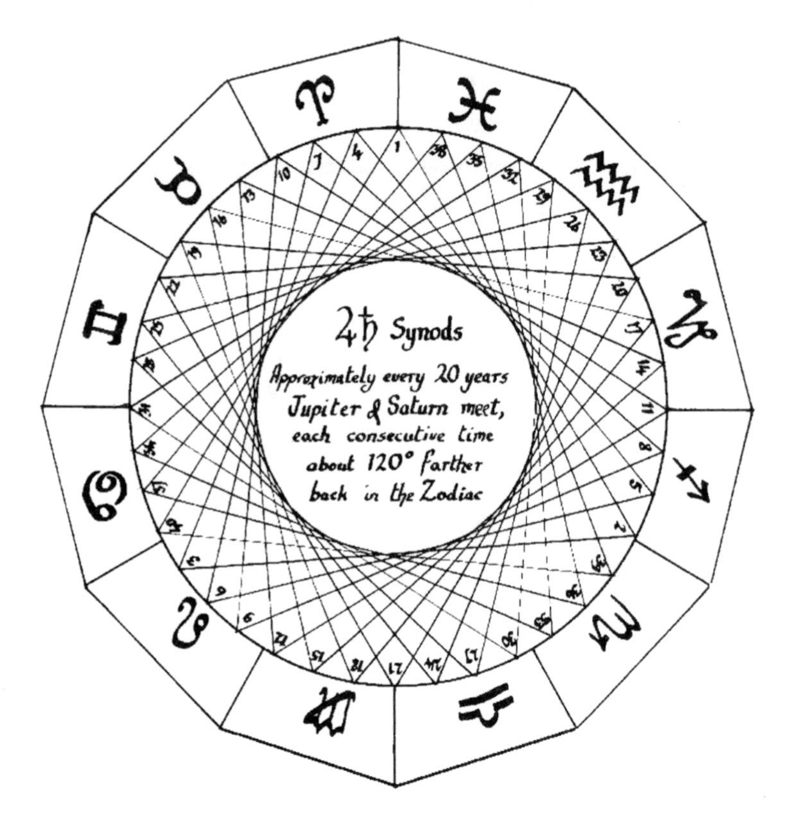

Figure 2.1. The Trigon Pattern of Jupiter–Saturn Conjunctions
The Trigon is so named because each synodic meeting of Jupiter and Saturn finds them approximately one third of a circle back from the previous synod. Three synods thus divide the zodiac into three nearly equal sections. The synod is approximately twenty years in duration and was well known to Mesopotamian astronomer/astrologers. Their most famous text, the Enuma Elish, *states that Marduk (Jupiter), their chief god, "bade the moon come forth, entrusted night to her, made her a creature of the dark, to measure time, and every month, unfailingly, adorned her with a crown." (Drawn by Robin Heath)*

TWO

ENTERING THE SYNODIC WORLD

A coincidence of planetary orbits is called a synodic meeting. Between such meetings, there are regular synodic periods. Such periods reveal the divine nature of the effects of the movement of planets upon Earth. When two synodic periods meet due to a shared numerical factor, a series of grand conjunctions occurs. The most celebrated of these is between Jupiter and Saturn, which coincide approximately every twenty years. Their conjunctions describe a series of triangles across the zodiac called the Trigon (see figure 2.1). When these triangles reoccur, each complete triangle is shifted nine degrees counterclockwise in relation to its predecessors.

Some synods are more common than others, such as the adjacency of the morning or evening star (Venus) to the crescent moon. Those who can read the sky see each combination as a landmark that speaks of the season in which it takes place. The star and crescent are always observed around sunrise or sunset, as illustrated in figure 2.2.

Some synods are irregular and rarely noticed. Mars has no simple synodic relationships to other planets, Venus and Saturn have regular patterns, and Jupiter with the Moon has a most subtle synod that integrates Mars to form a complete social calendar of the planets.

The Sun, and hence the year, is implicit in all synodic periods through the illumination of the full moon or the loops the outer planets form in the sky when opposite the Sun. Like the senior partner in a law firm, the Sun makes all this possible while appearing to do nothing.

DIVIDING UP GOD

The cycles of the gods in general, and Saturn in particular, cut up the zodiac into synodic pieces that are differentiated from each other through number. Not only do we travel through time within existence but also each of our moments can be known by the punctuation of the planets. The behavior of Venus around its maximum elongation from the Sun is unavoidably visible and always involves the crescent moon, either waxing in the western sky or waning in the eastern sky. Many national flags and mythologies are built on this periodic phenomenon.

Father Time, a.k.a. the Grim Reaper, often shown as a skeleton under a cloak with a sickle, is associated with death. The Egyptian god Set cuts up his brother, Osiris, and there appears to be a linguistic continuum among Saturn, Set, and Satan. An epic theme of tragedy is seen

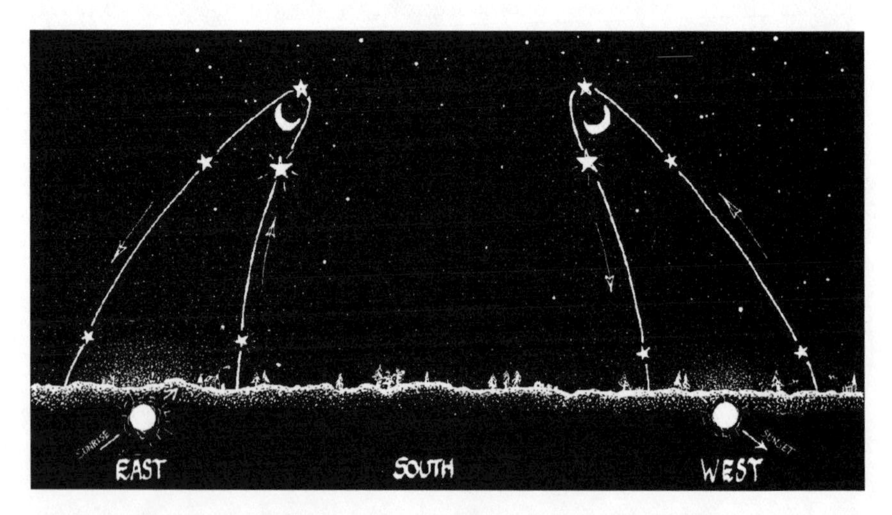

Figure 2.2. The Horns of Venus with a Crescent Moon (Drawn by Robin Heath)

in the world creation myth that involves the breaking of the cosmic egg or womb. This cosmic birth is the beginning of time for the creation. This breaking or cutting symbolism appears in circumcision, decapitation, castration, and the rituals that commemorate these acts. The creative act of birth involves the cutting of the umbilical cord—a rite of passage for a new life.

Figure 2.3 shows twenty-eight heads of corn growing from the body of Osiris. The mythologist Robert Graves writes of an ancient Osirian year as follows:

> That the Osirian year originally consisted of thirteen twenty-eight day months, with one day over, is suggested by the legendary length of Osiris's reign, namely twenty-eight years—years in mythology often stand for days, and days for years—and by the number of pieces in which he was torn by Set, namely thirteen apart from his phallus which stood for the extra day.

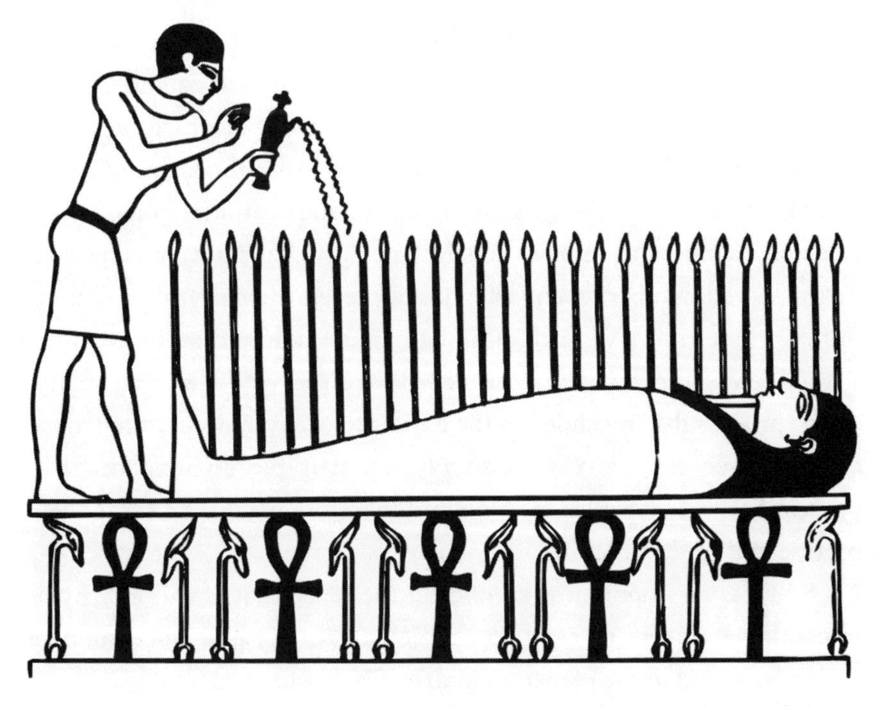

Figure 2.3. Corn grows from the body of Osiris. (Temple of Isis, Philae)

Symbolic cutting such as circumcision, decapitation, or castration is also invoked in the harvesting and milling of grain. The original sky god fell to Earth and became the crop of many grains whose generative parts are cut off, gathered, threshed, and milled. In the story of John Barleycorn, grain is boiled, fermented, and drunk as barley wine. In historical terms, the movement of the human race into agricultural societies coincides with a cultural awakening to time, a process memorialized in many mythical tales concerning cut-up sky gods.

Set (a.k.a. Saturn) cut up his sun-god brother, Osiris. The twenty-eight stalks in figure 2.3 represent that act of division, which gives the calendar its numerical characteristics. Before this, there were only the continuum of the stars and the crude measure of four seasons to the year—two solstices and two equinoxes.

DIVIDING UP THE YEAR

The number four is fundamental to the structure of the practical year. The Sun passes through spring and autumn midpoints and summer and winter extremes creating the two solstices and the two equinoxes. The number four emerges because of the axial tilt of Earth, which causes the Sun to appear to oscillate between north and south through the year.

The number twenty-eight multiplies the four cardinal points of the calendar (winter and summer solstice and spring and autumn equinox) by the sacred number seven. The number seven is construed as virginal in that it does not give birth, although it can be found in the creation. It represents something that existed before the world was created, a divine process that is echoed in the existence of seven notes in the musical scale (see chapter 8). Put simply, seven is the divine concept of process beyond or before creation and therefore is the first component represented in the creation of time.

Seven is an impractical number in the sense that it will not divide into other numbers to create a field of creation based upon number such as is found in the planetary matrix. Using our modern decimal system, when seven divides any number the recurring fractional part of

Twenty-eight-fold Division as Year, Month, and Day Divisions

If the creator (a.k.a. Saturn) wanted a perfect relationship to the year, there would be 364 and not 365 days in a year. The circle of the year could then be divided into twenty-eight periods of thirteen days—i.e., 364 days. Instead, the movement of the planet Saturn divides the year into twenty-eight periods of thirteen days plus one twenty-eighth of a day. Its synodic period is twenty-nine such periods long. The astronomies of ancient India and China divided the ecliptic (the zodiac) into twenty-eight and sometimes twenty-seven *nakshatras,* or lunar mansions.

This numerical arrangement may be usefully reciprocated within an annual calendar. Because the Moon moves thirteen degrees in a day, the same angular distance that Saturn moves in a year, it completes a circle of the sky in about twenty-eight days. A calendar of thirteen, twenty-eight-day months provides a whole number relationship between seven-day weeks, thirteen-week seasons, four-week months, and a 364-day year (364 = 2 x 2 x 7 x 13). Such a calendar was operative in ancient time, and until the 1950s was promoted as an alternative to the irrational Julian calendar.

The twenty-four hours we use for the day is based on the number six, not seven, through another Babylonian innovation. Similarly, the Babylonian division of a circle into 360 degrees conveniently allows the 365-day year-circle or the circle of the horizon to be divided by a huge range of numbers, as 360 = 2 x 2 x 2 x 3 x 3 x 5. Curiously, the Babylonian creation story, the *Enuma Elish*, remained based on seven days after the seven visible planets.

The outermost visible planet in the solar system, Saturn, undergoes twenty-eight retrograde loops in twenty-nine years. The outer Aubrey holes at Stonehenge contain fifty-six holes (2 x 28). This number is linked to lunar motion and eclipse prediction. Labyrinths are associated with time processes and are often sevenfold in their construction. The labyrinth at Chartres Cathedral, France, has an outer circle of 112 (28 x 4) markers. New relationships that project the story of the creation can be found within other parts of ancient monuments as discussed in chapter 9.

0.142857 results. Thus, one of the six numbers smaller than seven always begins its recurring fraction. For example, three sevenths is 0.428571.

Saturn is called "seven-rayed" in some traditions. He is considered evil and is rejected in a later creation myth that, having received the fundamental number seven, moves into the numerical system to evolve the creation with the numbers that work with each other.

It is worth considering the most enduring sevenfold legacy—the days of the week. The seventh day, or Sabbath, as the Israelites came to call it, is a day of rest. While the Christian world chose Sunday for its day off, Jewish tradition chose Saturday, Saturn's day. Although seven days might approximate the lunar month divided by four within a twenty-eight-day cycle, the lunation is actually 29.53 days long, so weeks have no astronomical foundation. Where did they come from? The Semitic religions developed a seven-day week after the captivity of the Jews in Babylon, while the Greeks obtained a more planetary version through their later conquest of cultures based upon the Babylonian system. The Hebrew creation story is poached from the Babylonian *Enuma Elish* and encrypts a sevenfold planetary and astronomic secret in the order of the days of the week (see figure 2.4).

The seven-day week, with each day named after one of the seven visible and mobile heavenly bodies, has become an apparently unjustifiable foundation for time on Earth, yet it reflects the nature of seven as the prime number, prior to what becomes time and number on Earth. Saturn represents that first type of time that underpinned the creation but could not do the creation work that other planetary numbers can.

SATURN—THE FATHER OF TIME

Saturn is the outermost planet visible to the naked eye from Earth. It is credited in myth with the creation of time itself at the expense of the starry sky, where unchanging patterns rule. The twenty-nine-year orbit of Saturn creates a large hour hand upon the zodiac. Thus, Saturn, or

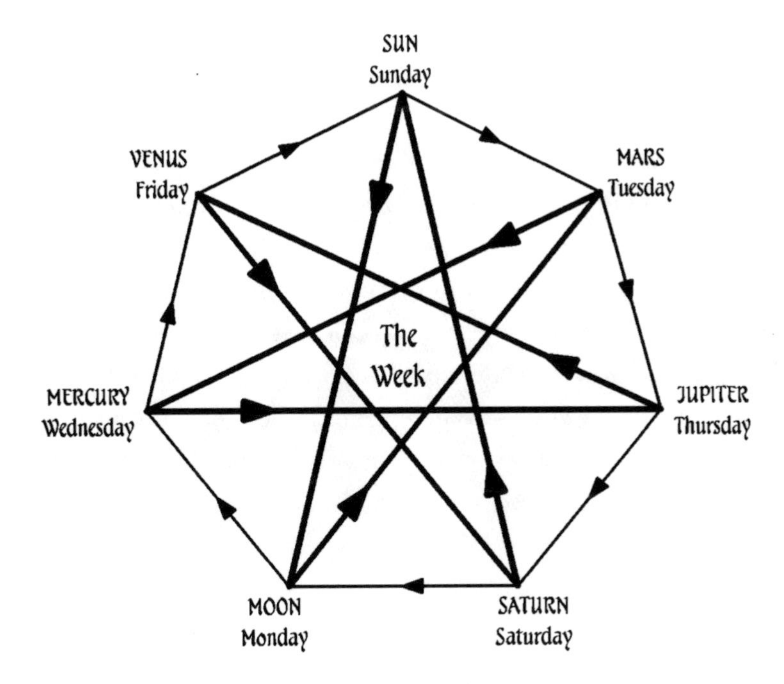

Figure 2.4. The Seven Heavenly Bodies
The "Chaldean Order" of the planets places them around a heptagon in order of increasing angular velocity, shown counterclockwise from Saturn. Each hour of the day was governed, in this order, by a planet, and each day was considered governed by the planet ruling its very first hour. The resulting order of the seven-day week can be traced out through the heptagram star, moving on by three planets every day since seven does not divide evenly into twenty-four but has a remainder of three. (Adapted from a diagram found in Pompeii)

Chronos, as he was known by the Greeks, symbolized time for humankind.

Everything in the sky relates to life on Earth and the practical year of 365 days. But something strange emerges when recording Saturn's synodic year of 378 days. The retrograde loop of Saturn, seen in the night sky opposite the Sun, repeats every 378 days. The Sun takes a whole year plus an additional part of a year to again oppose Saturn, because Saturn has moved on around the zodiac. Saturn's movement is not great, amounting to about thirteen degrees a year, and so the 378-day synodic year of Saturn is about thirteen days longer than the practical year. The synodic periods of Saturn therefore relate to life on Earth through a simple numerical arrangement. Twenty-eight Saturn synods equal twenty-nine practical years with 99.985% accuracy (one and a half days).

Saturn/Chronos, the Creator of Time, has defined the period within which a whole number of Earth rotations fit within the solar year of 365.242 days. Saturn has defined this practical year of 365 days with his own special number—twenty-eight—rather than fitting in to the solar year of Earth's actual orbital period of the Sun, 365.242199 days.

A calendar based upon the celestial movement of Saturn/Chronos, the mythological Creator of Time, is the numerical description of time on Earth.

PASSING THE GOLDEN MEASURE

The different streams of mythological thought can be compared to radio stations broadcasting across the millennia. They carry interesting though enigmatic news: The great ruler Jupiter deposed Saturn. Some say it was a violent coup but others say it was a natural development. Whatever the cause, the slow and careful Saturn is said to have passed the measure to Jupiter so that the creation could inherit Saturn's numerical foundations.

It is generally believed that the measure in question was the previously mentioned Trigon series of Jupiter–Saturn conjunctions in the sky. These are highly visible and occur approximately every twenty years. Every sixty years this conjunction returns to the same part of the starry sky, a longer periodicity than even the orbit of Saturn (see figure 2.1).

However, a much more important measure is passed from Saturn to Jupiter every year through the agency of the Sun. Suppose that the Sun is in opposition to the Saturn–Jupiter conjunction. After a further 365.242 days plus 12.848 days, the Sun will once again oppose Saturn. Meanwhile, Jupiter has moved on somewhat faster and has separated

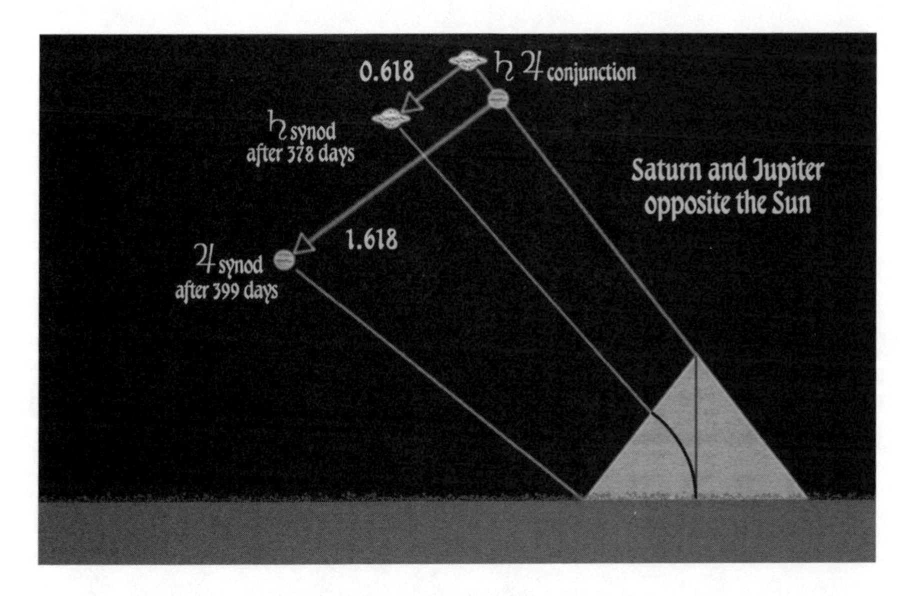

Figure 2.5. Phi in the Sky
The synodic periods of Jupiter and Saturn, 378 and 399 days, are numerically related via the practical year of 365 days to reveal the Golden Mean of 1:1.618.

from his conjunction with Saturn. An additional 20.78 days are required before the Sun opposes Jupiter, in a retrograde loop as observed from Earth.

The ratio of these two "catch-up" times, 12.848 and 20.78 days, is exactly in the proportion of 1 to 1.618 (accurate to 99.959%), which is that unique proportion called the Golden Mean or phi.

THE GOLDEN MEAN

This Golden Mean of 1:1.618034 is the proportion of most aesthetic appeal to the human eye. Mathematicians identify it with the Greek letter *phi.* When living structures are analyzed for proportion, they all reveal this ratio. Plants, skeletons, shells, the human face, and living forms in general embody it.

In fact, the Golden Mean arises in all recurring processes where

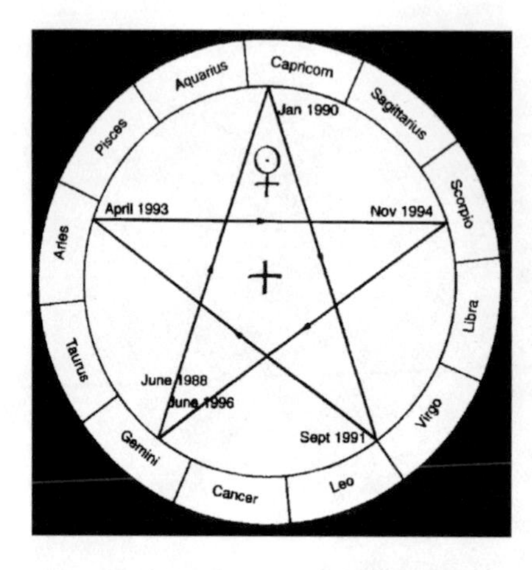

Venus's synodic period (Vsyn) = 584 days
Earth's practical year (PY) = 365 days
Venus's sidereal period (Vsid) = 225 days
Vsyn:PY = 1.6:1 (8/5)
PY:Vsid = 1.625:1 (13/8)

Figure 2.6. The Venus Pentagram, as Viewed from Earth
In eight years, Venus stands still in the sky five times. These standstills trace a pentagram star in the heavens. The period between standstills, 584 days, forms a close approximation to phi when compared with Earth's practical year, and was central to Mayan astronomical and religious practice. (Drawn by Robin Heath)

growth results from the current size and a past size of a structure. The Italian mathematician Leonardo Pisano, better known as Fibonacci (1170–1250), discovered this archetypal sequence of numbers in the twelfth century. Beginning with zero and one, Fibonacci added each two successive numbers to create the next number in the series. This generates a number sequence in which the ratio of adjacent numbers approximates with increasing accuracy the Golden Mean, phi—1:1.618034. Readers new to this idea will find an explanation in Table 2.1.

THE CREATION OF VENUS

This aspect of the Golden Mean is seen in the orbit of Venus around the Sun and the planet's synodic period as viewed from Earth. The orbit of Venus approximates the Golden Mean through a Fibonacci whole number ratio to the orbit of Earth of 8/13 or 0.615 solar year. In eight Earth years, there are five Venus loops or standstills against the background of the stars.

Figure 2.7. The Birth of Venus, *by Sandro Botticelli (1446–1510)*
"They say that Saturn cut off the private parts of father Ouranos, threw them into the sea and out of them Venus was born."
—MACROBIUS, A GREAT MEDIEVAL COMMENTATOR ON CLASSICAL MYTHOLOGY

As seen from Earth, the Venus synodic year repeats every 8/5 or 1.6 practical years to within 9.986% or about two hours. This enables us to divide the practical year into five units and Venus's synods into eight units—each exactly seventy-three days long.

Uranus (Ouranos) represents primordial and undifferentiated time. This existed before man noticed that Saturn cut the year into pieces that harmonize the practical year and the movement of Venus through the Fibonacci numbers 5, 8, and 13.

In myth and number, Venus and Saturn belong to the preliminary

Figure 2.8. Phi Fivefold Fun
The tracery of Venus's relationship with Earth is a pentagram. This view is as seen from above the axis of the solar system (i.e., the Sun is central). The scale is in millions of kilometers. (From Die Signatur der Sphaeren *by Hartmut Warm, courtesy Keplerstern Verlag, Hamburg)*

phase of numeric creation. Through Venus, five enters the earthly equation. This has more uses than Saturn's seven. Five includes the Fibonacci integer relations surrounding it and provides the root for the golden proportion found in the pentagram star.

The number six lies between five and seven, representing two multiplied by three. It unlocks the power of three, yet cannot be produced by Saturn or Venus. It takes Jupiter to introduce this new creation—an expanded numeric system—into planetary existence.

Table 2.1: The Fibonacci Sequence and Phi—The Creative Act

Imagine the most basic act of creation: starting with nothing and creating something. In numeric terms, begin with zero and create one. With these two numbers, we can evolve the basic time sequence by following the pattern of the Fibonacci sequence. Add the two numbers and the sequence begins to evolve: 0 + 1 = 1. The new number *one* is the result, or harvest, of the evolutionary process of creation.

Zero can be understood as the original state before the cosmos existed. One is the affirmation or impulse of the evolving universe. So, 0 + 1 = 1 is the first product of creation—the universe as a single whole.

Sequence	Time Zero	Time One	Time Two
Creation	0	1	1

If this process is continued, after time sequence three we have two aspects (1 + 1), God's will and the universe. Time four is 1 + 2 = 3, then 3 + 2 = 5, then 8, then 13, then 21, then 34, then 55, then 89, then 144, and so on. This is known as the Fibonacci sequence.

As the series progresses, the ratio between adjacent numbers in the sequence becomes closer and closer to the Golden Mean or phi—1.618033989 . . . Phi is found universally in processes of evolution, growth, and decay; the form of organic life; and the orbital periods and distances of the planets.

Mathematically, phi is expressed as (root 5 + 1)/2.

Geometrically, phi is obtained from the following construction:

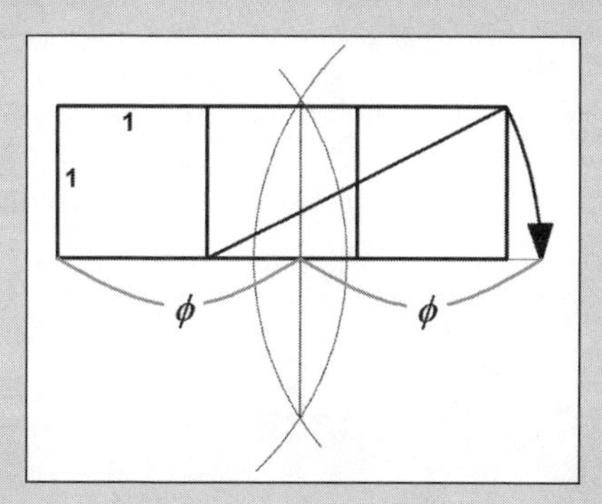

Cosmically, phi divides the whole into two parts, each in relation to the other according to the Golden Mean.

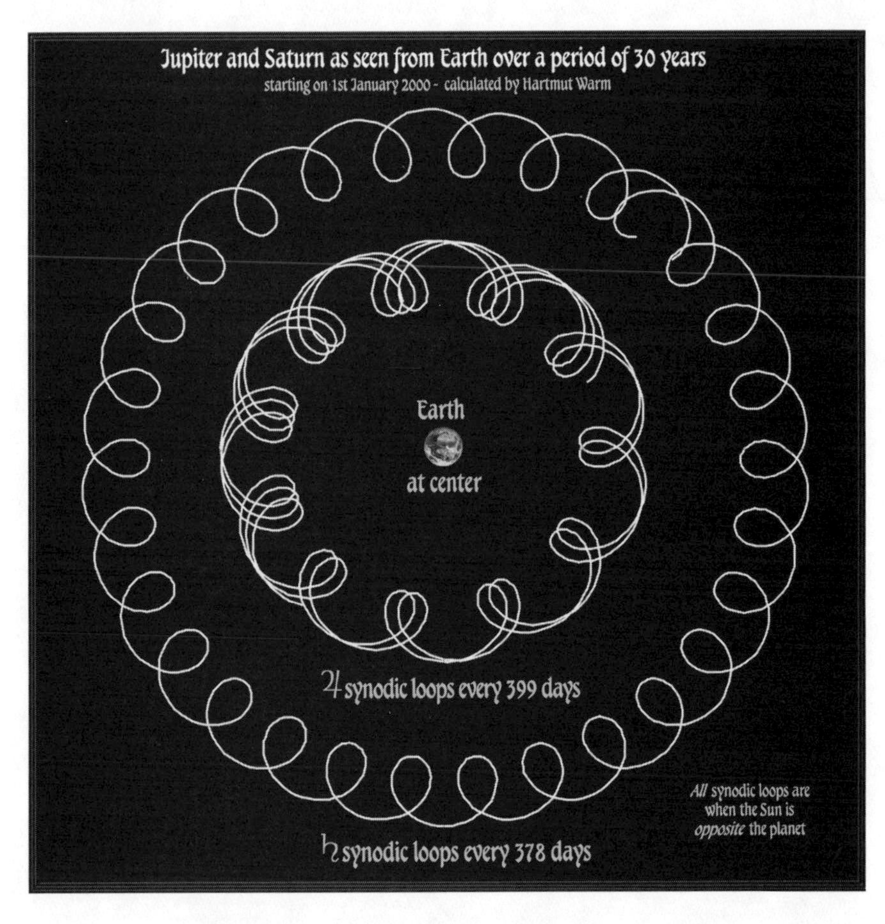

Figure 3.1. The Loops of Saturn and Jupiter
The two largest planets in the solar system—Saturn and Jupiter—complete each loop in 378 and 399 days, respectively. They complete their round of the heavens in twenty-nine and twelve practical years, a ratio of 12:5, to 99.3% precision (adapted from Hartmut Warm).

THREE

JUPITER AND EARTH'S MOON

Having received his golden staff or measure from Saturn, Jupiter develops his rule of Earth's practical year. This happens in an unexpected way.

Twenty-five of Jupiter's synodic periods of 398.88 days equal 27.321 practical years, as illustrated in figure 3.2. This is a wholly unexpected result, for the sidereal orbital period of the Moon, 27.321 days, is numerically identical! This curious truth manifests some important numerical coincidences between Jupiter and the Moon. One of these is that there are 365 lunar orbital periods (LOP) within twenty-five of

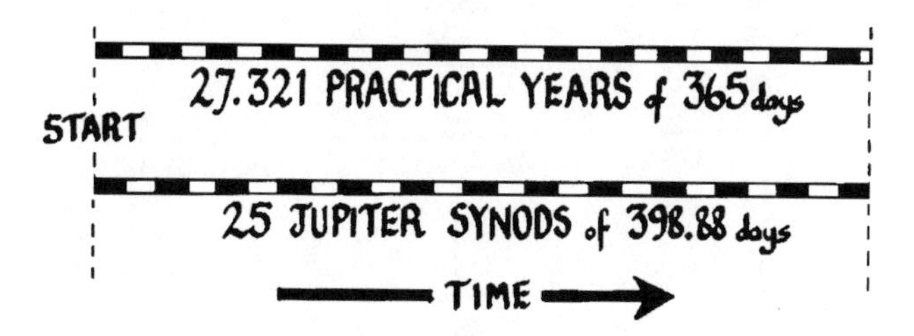

Figure 3.2. Relationship between Jupiter Synods and the Practical Year The same length of time is expressed as the product of two different sets of numbers: 27.321 practical years x 365 days per year = 25 Jupiter synods x 398.88 days per synod = 9972 days. Time runs from left to right. Both of the scales equal 9972 days. (Drawn by Robin Heath)

Jupiter's synods (to better than 99.996% accuracy). This is illustrated in figure 3.3, which shows that a single unit of time links different cosmic time cycles.

The workings of the demiurgic mind can be in reverse to our analytical thinking because it chooses from pure numeric relations prior to making them real. This gives the freedom of the cosmic Genius, a word derived from the Latin phrase meaning "to give birth." The numbers involved here form a narrative of the work of the demiurge.

The 365 sidereal lunar orbital months (lunar orbital periods or LOP) in twenty-five Jupiter synods confirm how many days there will be in a practical year, but they do this through the Moon, which orbits above Earth's rotation.

Since there are 365 LOP in twenty-five Jupiter synods, we can conclude that in one Jupiter synod there must be 365/25 LOP. This is a great key because it gives us a new convenient unit of one twenty-fifth of an LOP. We can then find the number of LOP/25 in a practical year—334. Later, we shall see that, by means of this unit, Jupiter

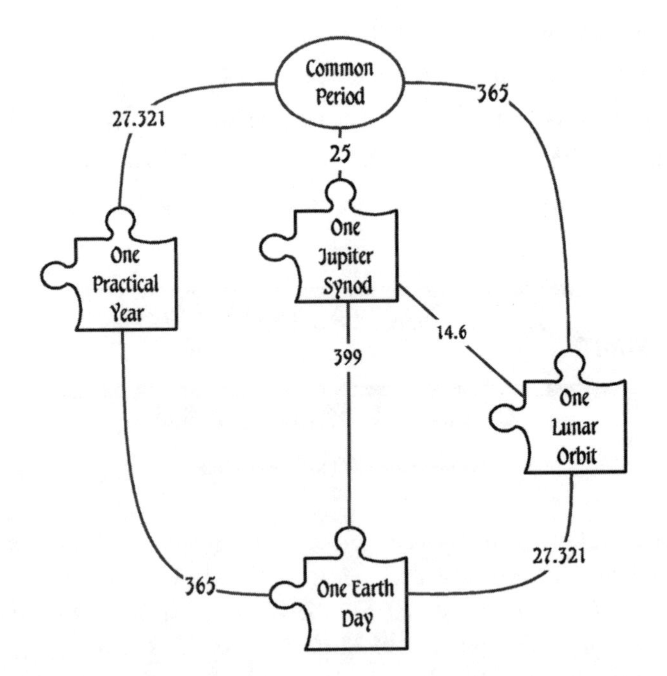

Figure 3.3. The Matrix for Twenty-Five Synods of Jupiter

installs the Golden Measure, which he received from Saturn, within the Earth-Moon system.

The fraction 365/25 equals 14.6. A fundamental unit of measure in Egypt was the *remen*, which is 14.58 inches, or 99.86% of 14.6. Is this an accident or did the Egyptians know about the planetary matrix? The remen records the synods of Jupiter while counting the number of lunar orbital periods within the same period. For this to be accomplished, the inch must have been passed down from ancient times. Chapter 7 discusses the mystery of the inch and its relation to the matrix.

The diagram below shows the relation of the Jupiter synod to the practical year. The Jupiter synod is 365 LOP/25 units long (399 days). This makes the practical year 334 of these units (365 days). 334/25 equals 13.36 (or 13 and 9/25), so there are 13.36 lunar orbital periods in a practical year. This corresponds with 99.995% accuracy to the actual lunar orbital period of 27.321 days.

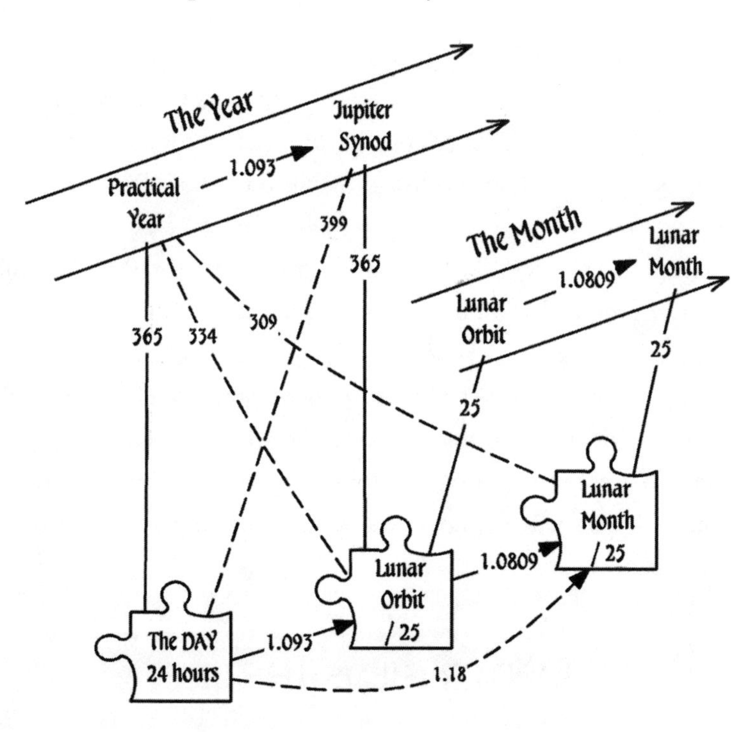

Figure 3.4. Proportions between Jupiter's Synod and a Practical Year Relative to the Moon

JUPITER ACHIEVES THE GOLDEN MEAN

The Sun moves almost once around the zodiac within a practical year. Consequently, the number of lunar months (lunations or new moons) in a practical year is one less than the number of sidereal lunar orbital periods (LOP). Thus, 13.36 minus 1 is 12.36 (which is 12 + 9/25). We may now equate the practical year as being 309/25 lunation periods of 29.53059 days or 309 LUN/25.

The number 309 is a very significant number. The Egyptians knew it as the number of lunations in twenty-five practical years and derived their calendar from these numbers. It is half of 618, an excellent approximation (99.9945%) of one hundred times the fractional part of the Golden Mean, which is 0.618034. The zero between the third and fifth decimal places in the Golden Mean makes it possible to approximate the fraction using only three decimal figures: 0.618.

The lunar month represents a combined relationship of the Sun, Moon, and Earth—a triad of periodicity that is greater than the sum of its parts. The phases of the Moon, from waxing to full moon and waning back to the new moon, have an average period of 29.53 days. Jupiter is synchronized to the Moon using units based on the number twenty-five (five squared). This Jupiter-Moon unit includes the lunation relationship via the Golden Mean as the number 309. This again involves the practical year, as 365 whole days, as it descends from the rule of Jupiter's synodic period, which is 365/25 lunar orbital periods.

One can only be dazzled at the elegance of this solution. Or perhaps we should call it a methodology! An interesting inference from this is that all the orbital motions of these three bodies must work within this solution, making them strangely related to each other through similar numerical factors.

JUPITER GILDS THE LILY

As seen in the last chapter, the practical year is divided into five equal pieces by the eightfold, Fibonacci-based synodic year of Venus.

Jupiter has innovated 309/25 lunations in a practical year because the number 309 is half of 618 (0.618 x 1000) and forms an excellent approximation to 0.618034 times 500. But five hundred divided by twenty-five is twenty. Therefore, there are twenty times 0.618 of a lunation in a practical year (20 x 18.25 days = 365 days).

Thus, a new unit has emerged that is 0.618 of the average lunar month (29.53 days). This unit is exactly 18.25 days long and is precisely 1/20 of a practical year. This numerical achievement, using the Moon, creates a more accurate Golden Mean timepiece than Venus.

These facts are reflected outside of the practical year since the reciprocal of the Golden Mean has the same fractional part as the Golden Mean itself. This is a unique property of phi. If the practical year is one unit, then each of its twenty pieces of 0.618 lunation when multiplied

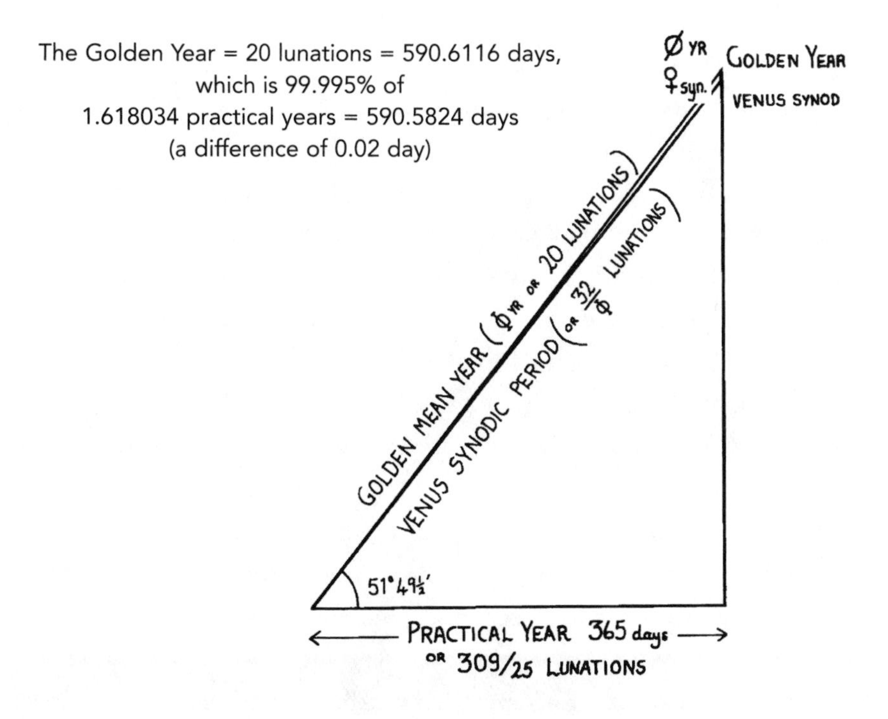

Figure 3.5. Golden Year Relationships among Twenty Lunations, the Venus Synod, and the Practical Year (309/25 Lunations) (Drawn by Robin Heath)

by the Golden Mean becomes twenty whole lunations within a Golden Year of 1.618 times the practical year.

This gives birth to a significant new type of year that overarches the practical year. To an accuracy of 99.995%, this Golden Year of 590.5824 days (phi multiplied by the practical year) has the same duration as do twenty lunations (590.6116 days).

This "Golden Year" is comparable to the Venus synodic year of 1.6 practical years (584 days). In creating new units of 18.25 days, of which the practical year contains twenty and the Venus synod contains thirty-two, an inner Golden Mean with the Moon complements the outer Fibonacci ratio of five to eight between Earth and Venus.

THE CHARIOT WITH ONE WHEEL

What really happens when Earth turns? The rotation of Earth describes periods that are measured in days. The solar year is 365.242 days long, the lunation period 29.53 days long, and so forth.

Earth orbits the Sun and, from Earth, the Sun appears to move through the stars. But the stars are lost in the brightness of the daytime skies and this obscures the Sun's progress from human view. However, through observation of the inexorable seasonal changes in the positions of the constellations, the Sun's motion can be determined.

The sidereal day is defined by the rotation of Earth relative to the stars. But this is different from what we commonly call a day, the full title of which is a *tropical day*. Our day includes extra time for Earth to catch up with the Sun before another sunrise. Our clocks are synchronized to this tropical day of twenty-four hours (1,440 minutes).

The Sun circumnavigates the zodiac in 365 tropical days, within which 366 sidereal days have occurred. There is one full Earth rotation more than there are sunrises within a year. This hidden oneness within the year is recapitulated in the one-unit difference between the number of sidereal days and the number of tropical days in a practical year.

The small catch-up time in every day is about three minutes and fifty-six seconds long. This unit defines not only a sidereal day with 365 such units but also the practical year of 365 tropical days. The catch-up unit is the difference between the duration of a sidereal day and that of a tropical day. It relates the Sun's daily motion to the rotation of Earth and is a fundamental unit of Earth time (figure 3.6).

Jupiter uses this three minutes and fifty-six seconds as part of a lunar timepiece to integrate the Golden Mean with the rotation and orbit of Earth within the practical year.

THE MOON GATHERS THE TEN THOUSAND WATERS

The sidereal day (the duration of one rotation of Earth) is a very significant cosmic unit. The Jupiter synodic period of 398.88 tropical days is within 99.993% of four hundred sidereal days long. Therefore, twenty-five Jupiter synods (365 lunar orbital periods) equal 10,000 sidereal days since four hundred times twenty-five is 10,000.

A sidereal day differs from a tropical day due to the motion of the Sun during one tropical day. The three-minute-and-fifty-six-second time difference between these two days, the aforementioned catch-up

Figure 3.6. A polar view of Earth's equator showing sunrises for two consecutive days. Compared with clock time, the stars rise three minutes and fifty-six seconds earlier each evening. (Drawn by Robin Heath)

unit, is quite useful when applied as the unit to measure the length of these days. A tropical day has 366 of these units while the sidereal day has 365 of the same units. The difference between the two is one unit.

Since 365 lunar orbits equal 10,000 sidereal days, it follows that a single lunar orbit has a duration of 10000/365 sidereal days. There are 365 units in a sidereal day, and therefore 10,000 units in a lunar orbit, so this new unit of time is 1/10000 of a lunar orbit. One ten-thousandth of a lunar orbit coincidentally is three minutes and fifty-six seconds in duration. The proportions in the Jupiter cycle combine with the lunar orbit, solar year, and Earth's rotation to generate a parallel number system involving the numbers 25, 40, 365, 366, 400, and 10,000.

This daily catch-up unit I shall a *chronon*. Its existence means that the rotation of Earth is synchronized with both the lunar orbit and the Jupiter synodic period using a time unit of about three minutes and fifty-six seconds.

The sidereal day of 365 chronons is the equivalent of the 365-day practical year, the chronon itself is equivalent to the sidereal day, and so on. The creation of equivalents through exact scaling enables a larger structure to be modeled within itself on a smaller scale. This is a recipe for the integration of sympathetic vibratory rhythms between the greater and the lesser structures, a planetary law of subsumption.

Earth resonates with Jupiter through the Moon. The slingshot effect of Jupiter's gravitational field propelled the meteor that collided with Earth and created the Moon. Therefore, the resonance between Jupiter and Earth via the Moon is probably of Jupiter's making. However, the orbit of the Moon took time to stabilize through the tidal interaction with water on Earth. So, we must ask if the present numerical arrangements are stable. If they are not, we live in an extraordinary epoch of numerical relationships (see chapter 9).

THE MYSTERY IN JUPITER'S ORBIT

The previously revealed unit of one twenty-fifth of one lunar orbital period (LOP/25) may also be found in orbit around Jupiter, confirming

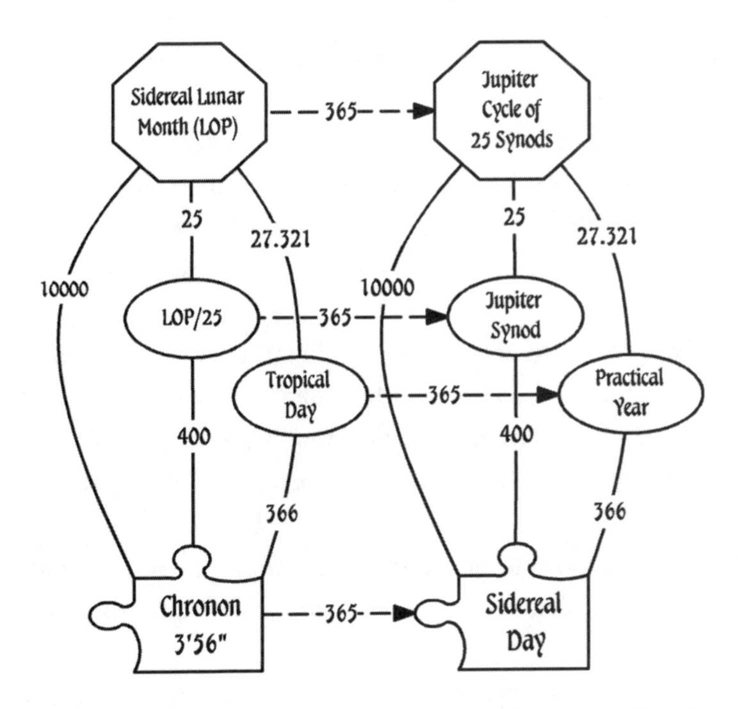

Figure 3.7. Parallel Lunar Orbit and Jupiter Lozenges Showing Their Resonant Parallelism

In 1/25 of the lunar orbital period, Earth rotates through the same angle that the Sun moves in one Jupiter synod, a period that is four hundred sidereal days in length. Since a Jupiter synod is four hundred sidereal days in length, there are 10,000 sidereal days in twenty-five synods of Jupiter. This creates a new unit of time, the chronon, which is 1/365 of a sidereal day.

the significance of this unit and establishing another Golden Mean relationship to the Moon.

Consider Arthur C. Clarke's *2001: A Space Odyssey,* where primates encounter a mathematically constructed monolith. This encounter triggers an evolutionary process that leads to the use of tools and technologies that enable the discovery of a similar monolith on the Moon. This in turn triggers a powerful transmission to an object orbiting Jupiter. There another monolith is found that is the connection to a great secret—the Star Gate. Could the monolith and the Star Gate symbolize the Golden Mean and the numerical understanding of the planetary system to open our eyes to the Cosmos in a new way?

It is amazing that Jupiter "created" a unit of time that is one lunar orbit divided by twenty-five. Yet this unit can be applied to the 1.769-day orbital period of Io, Jupiter's inner moon, within which are found 1.618679 such units. This approximates the Golden Mean to 1/25 of a percent [99.996%]!

The following basic facts emerge from this discovery:

- Twenty-five Io orbits equal phi lunar orbits.
- Since there are 334/25 or 13.36 lunar orbits in a practical year and 309/25 lunations in a practical year, we find that there are 334 Io orbits in twenty lunations or phi practical years (see figure 3.8).

The implication of these numeric relationships is that Jupiter exercises hands-on rule over time on Earth. In mythic terms, if the movement of Earth were a mill, Jupiter owns the mill and the Io–Moon relationship that invokes the Golden Mean is the handle of this mill.

Io is kneaded into volcanism by Jupiter's gravitational forces and by disturbances from the other satellites. It lies within a torus of charged particles or ions. These are pumped, by a flux tube that carries five million amps, from Jupiter's north pole, through Io, and then back to the south pole. Jupiter's magnetic field, if we could see it, would appear larger in the sky than our moon.

WHAT HAS BEEN ACHIEVED

The connection of the Moon to Jupiter and the practical year via 365 lunar orbits in twenty-five Jupiter synods calibrates the rotation of Earth to both these bodies. This is possible because the Jupiter synod is four hundred sidereal days long. The practical year of 365 tropical days must be 365 plus one sidereal days in length, and the resulting difference between a sidereal and a tropical day of 1/10000 of a lunar orbit is parallel to the relation that one sidereal day is 1/10000 of twenty-five Jupiter synods.

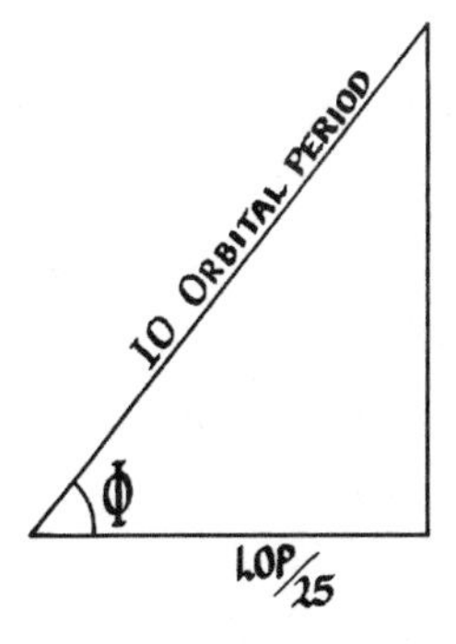

Figure 3.8. Io–Moon Triangle of Phi (Drawn by Robin Heath)

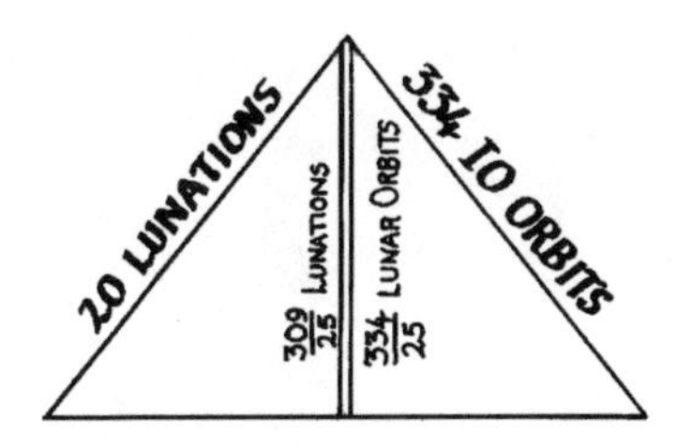

Figure 3.9. A Golden Mean "transformer," Io's numerical orbital link to twenty lunations or phi practical years. The height of this pyramid is 365 days. (Drawn by Robin Heath)

Because 10,000 (400 x 25) chronons are equivalent to one lunar orbit, the LOP/25 unit has to be four hundred chronons long. This represents the same angle as swept out by Jupiter in one synodic period. Everything has returned to the way it was measured when the Egyptians described Thoth, their ibis-headed moon deity, the god of wisdom, learning, and the arts. The etymology of the word *moon* runs deep and includes concepts of measure, suggesting the measure that was passed to Jupiter by Saturn. That measure, the Golden Mean, has been built into Earth's celestial view with an accuracy of 99.9945%. This is a great improvement over the Venus approximation, though Venus does create an excellent fivefold resonance within the zodiac, highly visible to earthbound observers.

Jupiter's work relates the sidereal rotation of Earth, the tropical day, the practical year, the orbital motion of the Moon, and the combined motion of the Sun and the Moon in the lunar month. Current cosmological orthodoxy suggests that these things should not be connected numerically in this way and are therefore displaying an inherent level of order in the cosmos. This fact directs us back to Earth and the unique life and biosphere that exist here.

FOUR

AS IT IS IN HEAVEN

The idea that planetary motions are a necessity for life on Earth draws attention to the fact that life is a vast display of interacting and evolving information. What is the purpose of all this information?

Cosmic unfolding applies simple principles to create an interesting world of existence. Creation is by definition productive and representative. It is associated with a maker or builder that the Greeks called Demiurge, the inventor of a creation mechanism that operates automatically on the energies of life.

Jupiter's title King of the Gods recognizes that he leads the forces of dark planetary matter within the Sun's aura. The abstract properties of number in planetary movements came into existence during the coalescence of the solar nebula in which vast bodies of dust and gas accreted through impact and disturbance, resulting in relationships that resonated using smaller numbers.

Just as the Milky Way galaxy created the solar nebula, so the dynamics of the early solar nebula led to long-term numerical solutions in orbital interrelationships. But, as seen already, the sidereal motions are translated by our subjective view from Earth, the viewpoint of life. This synodic world subtly interprets and re-projects planetary reality in a highly coupled connection to both Earth's rotation and its practical year, through the Moon.

The Moon is a translator or lens through which the motions of the planetary and starry worlds are registered. Though lifeless itself, it is involved with life but in a way different from the Sun, which provides the energy for life. The Moon represents the world of dark planetary matter and hence the magic of number in motion, which is seen in the forms life builds.

DO-IT-YOURSELF NUMERICAL ARCHITECTURE

When past civilizations turned to building, they identified their productions with the demiurge of celestial numerical form. The Golden Mean found in the Moon and Venus appears at the heart of creating the world of form itself. Schwaller de Lubicz identifies the Golden Mean as "the fundamental scission," or division of one into two, that creates three things—the original whole and two parts, one in golden proportion to the whole and the other in golden proportion to that.

Civilizations respond to these inherent numerical signatures and base their culture around the numerical architecture of calendars or seven-day weeks. Ancient sacred buildings reflect themes that are explicitly astronomical and numerical. The priests who built and maintained them represented the demiurgic god. The observation of patterns in the sky and the desire to elucidate their numerical structure served as the genesis point of these buildings.

John Michell and Christine Rhone take up this theme in their remarkable book, *Twelve-Tribe Nations and the Science of Enchanting the Landscape*. They write, "Numbers were not invented but discovered. They existed as unmanifest archetypes before there was anything for them to quantify. When there were no three objects, nor four or five of anything, the numbers 3, 4, and 5 were already immanent, and when creation took place, they were available to give it form and proportion. This perception gave rise to the creation myth behind ancient philosophy, that the Creator's originating thought was of a perfectly harmonious code of number, and from that pattern developed all the forces and phenomena of nature."

The authors demonstrate convincingly that the ancient universal adoption of certain numerical signatures and rules enabled civilizations to live in a state they call "enchantment." This was brought about through the numerical structures of calendars, weights and measures, land allocation, and monumental architecture. Simple fractions can transpose units of length from one system to another, incorporating key constants in the dimensions of Earth. This point is elucidated convincingly in J. F. Neal's book, *All Done with Mirrors*.

Why did so many different cultures build such great monuments, enshrining their numerical and astronomical systems? Although ancient monuments serve as informative records allowing modern researchers to read their numerical messages, these buildings must have had an immediate use for those who built them. In some instances, this application appears to have changed over time. The Egyptians were quite willing to tear down one massive temple and build a new design, often many miles away, using some of the original stonework. The Stonehenge bluestones were not dragged hundreds of miles from Wales to Wiltshire on a mere whim.

Ancient monuments are cosmological structures on a par with our particle accelerators and radio telescopes but with the focus on numerical, rather than atomic, reality. The numerical content of stone circles, for example, is extremely sophisticated. They may have had only a few observational uses, but many numerical relationships were captured within them.

The most famous megalithic monument is the Great Pyramid at Giza in Egypt. Its slope angle is the Golden Mean ratio, and its dimensions enabled John Greaves to establish the length of the royal cubit—1.718 feet. The southern base length is 756 feet, or 440 of these cubits, making the height 481 feet or 280 royal cubits. The product of the two dimensions, in any units, produces the length of the degree (68.88 miles) at latitude thirty-one degrees north. The dimensions of the Pyramid thus relate it to its location on Earth, while its inner passages date its construction through their alignment with key fixed stars.

Another great work is the Temple of Luxor, also in Egypt. Schwaller de Lubicz calls this the Temple of Man, in which cosmogenesis is mapped onto the human body. But, one might ask, how can the human body represent the cosmic process and therefore have astronomical and numerical structures? Perhaps the simplest answer to this is that man represents the most evolved result of biospheric evolution to date.

IN THE MIND OF MAN

Medieval thinkers believed that parts of the body corresponded with particular planets or constellations of the zodiac. Was this concept a part of an earlier ancient wisdom? Imagine if the worlds of human concept and sensation were separated by the nature and mechanics of our physical bodies. Then, when old concepts could no longer connect to the world of the senses because there was no longer any understanding, old wisdom would be lost. It could then be recovered only when that understanding returned.

This idea describes an ancient doctrine which predicts that a dark age would ensue when mankind would forget the truth entirely and terrible events would occur. The cause of this dark age was read in the sky as the precessional cycle—the changing orientation of the north pole to the Galaxy. The ancients saw world history emerging out of the sky just as creation had done. The sky was the higher world in a very literal sense and man was reimagining the cosmic process.

The different types of knowledge that have survived can now be seen as parts of a single worldview. The different component formats included mythology; numerical architecture reflecting the sky, Earth, humanity; and energy practices found in shamanism and other lineage-preserving orders. The secrecy of these practices saved them from obliteration and dilution, albeit temporarily.

Christians saw stone circles as evil pagan relics and often destroyed them. Muslims still destroy sacred art, apparently not knowing that the numerical formulas inherent within them are identical to those found in sacred Islamic art. At a cultural level, the present situation is bad for mankind because we have no integration with the higher world.

THE ENERGIES OF LIFE

One much maligned system of understanding is that of the four elements. These are in fact the four energies used only by living systems—constructive energy, vital energy, automatic energy, and sensitive energy—and are similar to the four more familiar energies of material systems—heat, force, chemistry, and flexibility. J. G. Bennett mapped these energies in his epic works, especially *Energies: Material, Vital & Cosmic.*

The living energies coexist with the related material energies. The body has heat, muscles, bones, and soft tissues and also employs constructive energy, the first life energy, to build its replicating chemical factory. But constructive energy lacks independent direction, so vital energy, the second life energy, provides integration and regulation on the edge of conscious awareness.

These first two life energies correspond with earth and water. Immediately these designations help our understanding, since the interconnection of the familiar with the otherwise mysterious and invisible life energies is an act of cosmogenesis. Vital energy operates like water, moving Earth around in its geological cycle and preparing it for life. This understanding has no beginning or end and belongs to a spiritual science that integrates complexity in an inspiring and simple way.

The next life energy, automatic energy, is like air and is often seen as mechanical thought. When something new is learned, the body can do it automatically. This happens so quickly that most of normal life is based upon automatic energy as we drive cars, enter familiar rooms, and need very little observation to perform routine tasks safely. But mechanical energy is dangerous, like a machine with no driver, so the next life energy is sensitive energy, which corresponds to fire.

The basic sensitivity provided by the senses ignores some important things. We know that the visual cortex does an enormous job, below the level of consciousness, to allow simple control of focused activities. The detail is filtered out. However, under the right conditions some small fact is sometimes seen out of proportion, such as someone's face in a crowd or an object that could be useful. This ability is often useful to solve a problem or to recognize a dream sequence that prepares the mind. When this occurs, the energies of the sensory world and those of the brain are working with an intelligence that mechanical thought cannot achieve, for "Having eyes [mechanical energy] see ye not." This is fire, associated with the powers of the subconscious mind but requiring the highest of life energies—sensitive energy. Sensitive energy is like light since it allows one to really see.

This treatment of energies allows mythology to reveal information through the circumstances of stories. It creates a technical language through which energy transformations can be communicated by intermediaries who enjoy a good story.

ENERGY PRACTICES

There are also four cosmic energies. Though the three higher ones are largely hypothetical for human existence, the lowest is called conscious energy, even though it is beyond the ordinary experience we call consciousness. Conscious energy integrates the living energies while independent of life itself. As quintessence, it is a fifth element like ether or mind and is a product of the planetary world. While sensitivity is the correspondence between a person's outer and inner lives, conscious energy is the understanding of life and the knowledge brought by such understanding.

Man's transformation to enter the cosmic process is tied up with the transmission of conscious energy from the sky. All mythologies describe this as the godly teaching of the arts of civilization and spiritual truth to mankind.

The life energies are sandwiched between the cosmic and the material realms. They form a bridge where cosmic energies descend to Earth and material energies ascend into civilization. This is the real purpose of life and of the biosphere, to mediate between the cosmic process and the material planet. The energy bridge creates high civilizations separated by dark ages—ages that allow the identity of the former civilization to be destroyed so that a fresh one can arise.

The different levels of energy possible lead to an energy meritocracy where stages of human evolution beyond the ordinary are achievable. Special individuals can interact with the cosmos through conscious energy and the creative energy above it, to bring about the extraordinary or even generate entire civilizations. So, prophets and messengers, shamans and priests, all are objectively judged by their capacity to productively integrate the higher living energies and lower cosmic energies.

SACRIFICES TO THE GODS

Earlier, the sacrifice of the starry sky god was shown to have created time. Sacrifice is a complex tradition seen too often in the form of human sacrifice, such as decapitation or castration. The sacrificial moment marks the descent and ascent of energy. Descending energy can be used to do work, seen in solar-powered homes and vehicles, the internal combustion engine, and other modern achievements. At certain times, special results can be achieved through such transference of energy. This is most true in the inner lives of men and women. The esoteric art of alchemy contained the belief that action at the right time and place can bring about energy transformations.

The outer ritual of sacrifice is partly a degenerate expression of a chilling fact expressed by Gurdjieff: Life must support the cosmos, especially the Moon. If the right energies cannot be produced by the biosphere, then lower energies in much larger quantities are demanded by the system through destructive events such as war and other catastrophes. In other words, ignorance of these energy systems is deadly and also inefficient. The cosmos needs evolution to become conscious, generating higher energies in exchange for human development. But new faculties are more and more hazardous since the human mind can easily become an evolutionary cul-de-sac.

SENSITIVE ENERGY AND THE BRAIN

Evolutionary biologists recognize two different characteristics associated with brain development in mammals. The first, called intelligence, is the capacity to comprehend relationships. The second, which is uniquely human, is cognition, in which symbols can take on abstract meanings, as found in the faculty of speech.

The enormous size of the human brain, while a liability for ordinary survival and selection, has enabled us to become very articulate. The aggregation of cognitive material—thoughts, stories, information—represents the after-effects of sensitive experience and hence of sensitive energy. Thus, cognitive brain development appears to have adapted to transform information in the biosphere beyond the automatic recognition of relationships in the environment.

This all corresponds with myths in which our progenitors confer the power of speech, which "even the angels did not have," upon humanity. With speech came the power to name things, an ability that was considered godlike and ultimately led to original sin. By speaking freely, humans created a world somewhat independent of the "biologic" of environment, and consequently a whole new range of behavior associated with "sin" emerged.

The geneticist Richard Dawkins has identified the evolutionary tendency in human genes to evolve the brain to have just these capacities of descriptive power. He goes on to suggest that the thought structures in such a brain tend to have a life of their own and that they replicate like genes. Dawkins named these replicating thoughts *memes*. Though born of genetic preferences, memes can often act in ways that contradict survival itself. This is seen in individuals who sacrifice themselves for a principle, enter monasteries and never breed, and even commit suicide at the suggestion of a leader (as discussed in *The Meme Machine*, by Susan Blackmore).

While genes are replicating, they must have a biosphere of vital energy to evolve the necessities of survival. In the same way, if memes are replicating, they too must evolve to suit the necessities of a higher world. In this case, the higher world is the noosphere, which is made of the highest life energy—sensitive energy—since memes are made of the automatic energy of ordinary thoughts.

Above this sensitive realm of experience we can now place the planetary matrix, a cosmic replicator of eternal numerical recurrence. Interestingly, the memes of ancient peoples, including their myths, appear to have been about the cosmic events of the matrix, rather than celebrities, special offers, products, and tourist destinations. The quality of the memes in our society and the mechanisms of their use are important areas for future research into objective ethics.

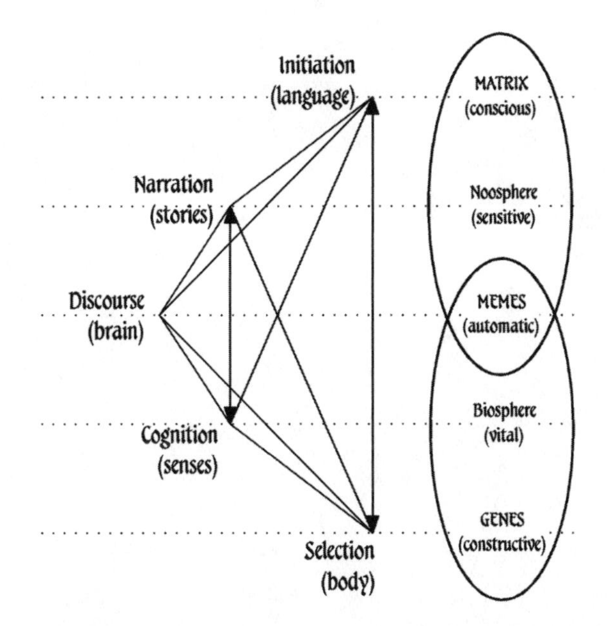

Figure 4.1. The Evolution of Information as Energies
A pentad is a five-term system based on J. G. Bennett's theory of systematics, which says that numbers form patterns of understanding within well-formed categories of meaning. These are found in the five levels of energy seen here, three of which support self-replicating behaviors.

The quintessence of memes is discourse, *through which they are formulated and reproduce. The range or limits of discourse are between* cognition, *(evolved by* selection *within the genes) and* narration *(evolved because life is a journey through time).*

The correspondence of initiation *with the calendar and a cosmic event is entirely traditional. It is not surprising, then, that mechanical recitation and ritual are the formats for such discourse and the narration of the stories found in myth. The greatest sacred journey for us all is the initiation into the meaning of our own journey: You either get it or you don't!*

Figure 5.1. Hephaistos returns to Mount Olympus with his smith tools of hammer and tongs on a donkey, for he is traditionally lame. He is returning so that his mother, queen of the gods, Hera, might be freed from a golden throne he made. (The Kleophon Painter, Skyphos, *side A, detail of Hephaistos, courtesy of the Toledo Museum of Art)*

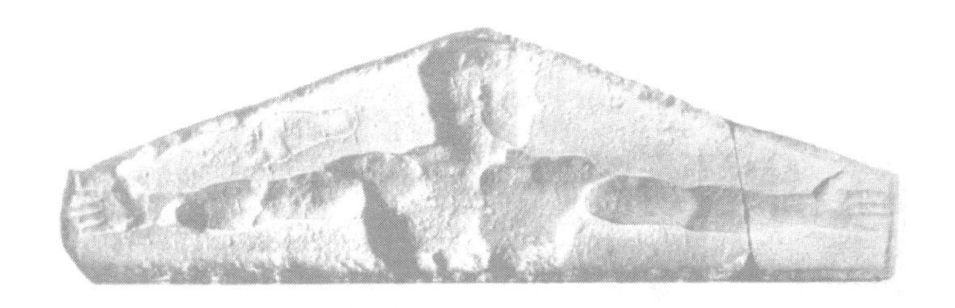

FIVE

MAKING THE GOLDEN THRONE

Few people realize that myth is a technical language placed within a popular format. While filled with sex and violence, impossibility and unlikelihood, it is through these allegorical energies that a rigorous technical content is maintained within an enduring oral tradition.

The interpretation of myths and their themes have been the subjects of many arguments. Today, psychological matters are the most popular candidates. Interpretations as archaic astronomical observation and speculation have intellectual appeal, and the view that they are the work of juvenile, superstitious minds thankfully is receding. Unique variations of similar tales have been found worldwide, making the cultural roots of myth a titanic secret, lost in time but glimpsed through the myths themselves.

If different levels of the world order and the biosphere correspond, it is entirely possible for astronomy and psychology to be referred to by the same myth. The task in this chapter is to discover the astronomical content of myth and find out if it corresponds in any way to the planetary matrix.

The classic book *Hamlet's Mill,* by de Santillana and von Dechend, gives many suggestions as to how astronomy is presented in myth. Its thesis focuses upon the phenomenon of Earth's 26,000-year cycle of polar precession and its consequent effect on world history. However, the matrix is a planetary phenomenon with elements operating down to minutes of time. It is obvious that any archaic astronomy referred to in myth was more than a calendar for long-term change. Therefore, we can see that precession is better understood as the larger backdrop for what has been revealed concerning the shorter periodic behavior of planetary motion—the planetary matrix.

It is clear that Chronos, who "is the head of the whole genus of Titans from which originates the division of things," is Saturn "giving from above the principles of intelligence to the Demiurge [Zeus/Jupiter] and [presiding] over the whole creation" (Proclus). The technical language in this statement involves division, or cutting up, similar to the emasculation of Ouranos to create Venus/Aphrodite or the division of Osiris by Set.

Order is being created out of the "disorder" of no numerical order within the physical cosmos, and, effectively, there can be no time within such chaos. Time can be known only through physical processes, and these must be material and living processes. Humans can experience time, as can animals to a lesser extent—a fact apparent in the faculty of memory.

So, the word *chaos* is a technical term for the universe before number, before the creation of the current planetary system and man's discovery of it. Without consciousness of time, there can be no time. The yogis of India describe the senses of man as the Indriyas—belonging to Indra, who is one of their Zeus personalities, a demiurge. Therefore, the organization of the human senses corresponds to the world created by Jupiter, and the biosphere evolved from a world within Jupiter's matrix of numerical activity.

Perhaps for every outer fact there is an equal and complementary inner fact, making for an unscientific process of discovery in which, though proof is after the facts, creation is before the facts.

A MATRIX MYTH

An orbit, in the numerical creation, can be termed the "wife" or "son" of a planetary god. Thus, Hera, the long-suffering partner of Zeus, can be considered the extended present moment of the recurrence, in eternity, of Jupiter. Human eyes see Jupiter at one point in its orbit, but to a being with a greater present moment, their own presence extends more broadly. In human experience, too, there are extended present moments in which competence requires keeping in mind factors outside current experience in a pattern of will.

If Hera represents Jupiter's orbit, among other things, then the fact that both she and Zeus have separate but complementary children who emerge directly from them, like virgin births, has new meaning.

Athena, the patron of the city of Athens, came out of the head of Zeus after he had swallowed the Titan Metis, whose name means "cunning intelligence," because his children otherwise might have threatened Zeus's creation. Athena is considered a moon goddess and, coming from Jupiter's head, is a perfect representative of the relationship of the Moon to Jupiter's synodic period.

Figure 5.2. The Birth of Athena from Zeus's head

A similar unnatural child, named Hephaistos, emerges from Hera (the orbit of Jupiter). He is born ugly and lame. Hera rejects him and drops him from Mount Olympus (Earth) into the ocean that surrounds it (the lunar orbit). Hephaistos represents the orbit of the Moon, not the luminary herself. He is the lunar orbital period divided by twenty-five, a number that is initially lame because it does nothing until its ramifications are consolidated.

The story of Hephaistos is a timeline that describes, in the technical language of myth, the simultaneous relationships within the planetary matrix. His orbital role representing Jupiter becomes dominant and transforms the lunar orbit into the Golden Mean connecting the Moon to Jupiter through lunation phenomena.

HEPHAISTOS'S JOURNEY

When Hephaistos lands in the ocean, he is initially nurtured by sea nymphs in an underwater grotto. There he sets up his first smithy, for he becomes the smith-god. Nine years pass (three times three (3 x 3) years, linking Hephaistos numerically with the Moon, the "triple goddess"), during which Hephaistos becomes an excellent smith, especially with gold. He makes a golden throne (figure 5.1) that, when sat upon, binds the sitter with invisible cords, and Hera is attracted to sit upon this golden throne.

Hephaistos's sojourn in the ocean depths produced something attractive that binds Jupiter's orbit to that of the Moon, creating an inescapable golden throne. What better invisible bonds can there be than those of numerical coincidence?

The grotto in the depths might be more fundamental than the lunar orbit divided by twenty-five. It could be the chronon, a suitable source for something golden, of which there are 10,000 in a lunar orbit. The number 10,000 is made up of twenty-five times twenty times twenty (25 x 20 x 20). Though Hephaistos's unit (LOP/25) is four hundred chronons long (20 x 20), there is another unit that becomes useful in smithing the gold. This is the orbital period divided by twenty, within which there are five hundred chronons (25 x 20).

While there are 20 x 500 chronons in a lunar orbit, there are 21.618 x 500 chronons in a lunation (10,809). The Golden Mean has arrived and, through "invisible bonds," binds the Moon to a "golden throne" (figure 5.3). Also, 1,080 (which is quite close to 10,809 divided by 10) is the numerical signature of the Moon in sacred geometry and biblical gematria. Remarkably, the Moon's radius is almost exactly 1,080 miles.

FORGING THE GOLDEN THRONE

It is a strange property of the matrix that two numbers derived from a common factor and two respective multipliers can coalesce back to a new common factor by multiplying each with the other respective multiplier. Obviously, any factor multiplied by the same two numbers must produce the same result! What must be appreciated in this relationship is that the two different intermediate results can exist in apparent independence while both have a factor in common and are common factors in a greater result.

Here is a simple example of this to illustrate what is happening that will also be useful later. Multiply the number one separately by both three and four, and of course you obtain the numbers three and four. But both three and four can only divide into twelve, and hence three times four (3 x 4) and four times three (4 x 3) have the same result: twelve. Therefore, any two time periods in a 3:4 ratio will inevitably have a common synodic repetition of twelve units and share a common factor unit with a length of one.

In the above example, LOP/20 links to the lunar orbital period through twenty and the lunation period through 21.618. Therefore, we can multiply the lunation by twenty to get the same period as 21.618 times the lunar orbit. The latter operation also shows us how many practical years pass during twenty lunations, 1.618—the Golden Mean—in what can be called a Golden Year.

The left-hand branch of figure 5.3 consists of multiples of 20:21.618. Therefore, it is the candidate for the golden throne since it

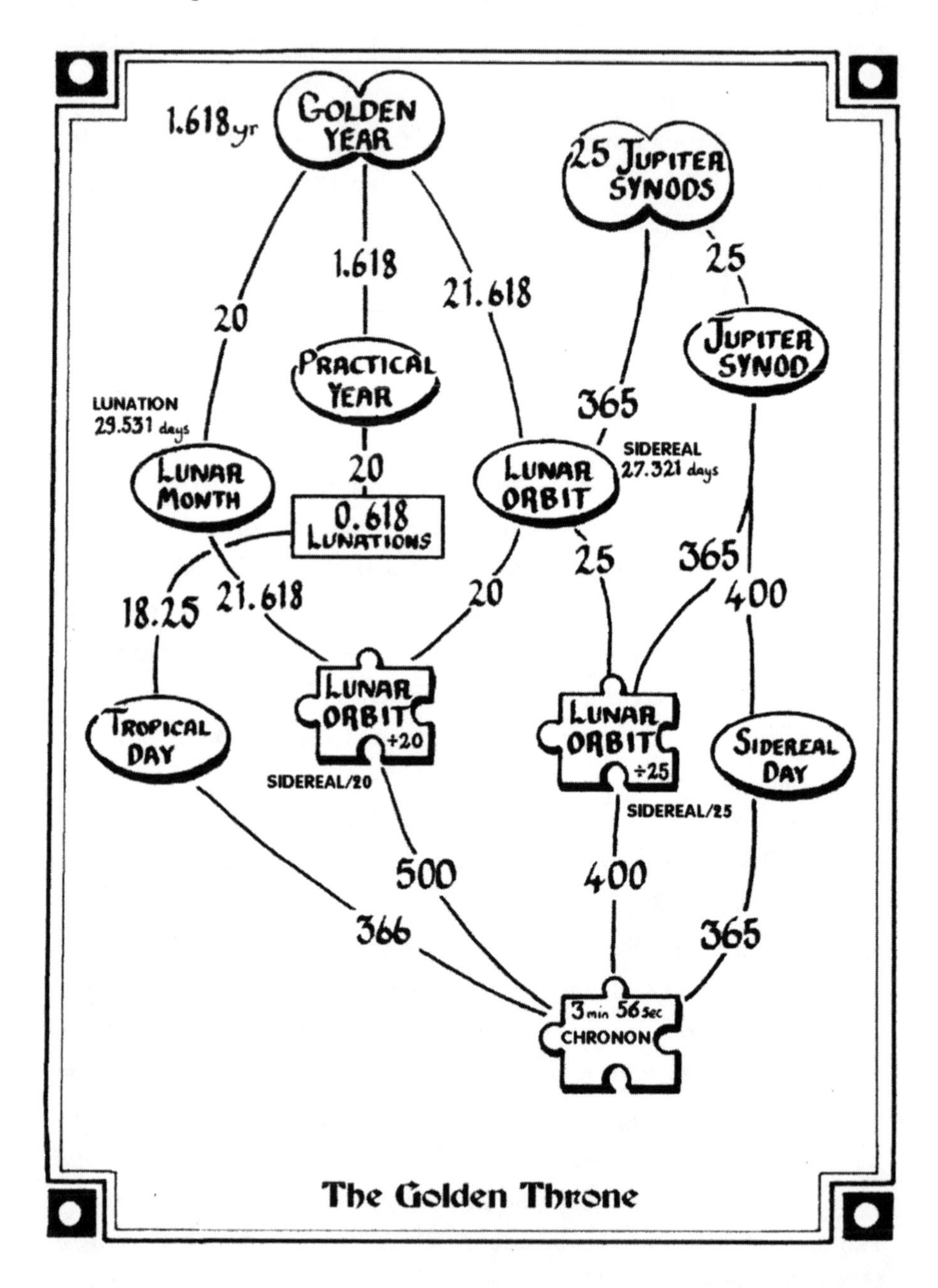

Figure 5.3. How Hephaistos forged a golden throne from Jupiter's year, the lunar orbit, and Earth's rotation. The chronon is the difference in time between the tropical and the sidereal day. (Drawn by Robin Heath)

defines the numerical space within which Hephaistos created this great work from his forge deep in the chronic ocean of time.

WHY THE GOLDEN MEAN IS GOLDEN

Hephaistos's golden throne is built through scaling, a technique that equates things of different levels or scales. More scaling reveals some new mythical relationships for this throne.

If there are twenty lunations in a Golden Year, then in a single practical year there must be twenty times 0.618 lunation. What happens to translate the lunation cycle into a unit allied to the Golden Mean?

In 0.618 of a lunation the Moon moves through just over two thirds of its orbit around Earth (0.668). It is self-evident that the Sun must move one twentieth of its yearly round in one twentieth of a practical year. An interesting fact emerges from this: The Golden Mean plus one twentieth equals nearly two thirds. (*Note:* The active part of the Golden Mean is its fractional part, namely 0.618. This is the reciprocal of the usual 1.618 Golden Mean. The more significant part is 0.618 because, in division of the whole, everything is less than one.)

The goldenness of the Golden Mean is revealed as the Sun's motion during two thirds of a lunar orbit. This is the most likely origin of its gold, since the Moon is associated with the metal silver. A truly beautiful higher result is seen using a forge whose fire lies at the center of the planetary system. Hephaistos used the very same metal so often associated with the Sun and hammered it into jewelry and ornaments out of numerical artifice. The resulting golden relationships were founded upon the orbit of Jupiter, the master Demiurge.

OF RAPE, MARRIAGE, AND ADULTERY

Sexual politics are at the top of the agenda in myth. The coitus here, though, is an act of numerical interaction usually based upon the diamond-shaped factoring of the planetary matrix.

Continuing in the myth of Hephaistos, Hephaistos marries

Aphrodite/Venus, either by his or his mother's demand. He is local boy made good, so why not offer him the hand of Venus—except, of course, he is not as good looking as she is. Interestingly, recent research has shown that a beautiful face is a face with many Golden Mean proportions within its features.

The new factor of 0.618 lunation, bestowed upon the Moon, is 18.25 Earth days long. Four of these make up seventy-three days, which in turn is one fifth of the practical year of 365 days. A Venus synod is eight such periods long or 584 days between recurrences of Venus in the skies above Earth, where Venus is observed as an evening or a morning star.

The 5:8 ratio between the practical year and the Venus synod has already been noted as part of the Fibonacci series, which approximates the Golden Mean, though poorly, as 1.6 practical years. The Golden Year of 1.618 practical years means that the Venus synod is now placed within the golden throne of twenty lunations built by Hephaistos. Also, there are twenty units of 0.618 lunation in a practical year and thirty-two in a Venus synod.

In this way, Hephaistos "marries" Aphrodite (Venus). This is a grand reunion of forces that were set apart when Chronos/Saturn castrated Ouranos and cast his genitals into the ocean foam to create Aphrodite, while giving the Golden Measure to Zeus/Jupiter. Zeus, in the form of his wife's offspring, becomes an orbital unit for the Moon, based upon Jupiter's orbit, and has finally married Aphrodite to the lunar orbit through a Golden Mean lunation unit.

Myth is filled with incest, parthenogenesis, and ethical ambiguity because its stories are not about ordinary people but are instead about gods with planetary identities, orbital interactions, and numerical desires. The last part of the Hephaistos story concerns Aphrodite's dalliance and subsequent adultery with Ares/Mars, the god of war.

Figure 6.1. Mantegna's Parnassus

The nine Muses dance below Ares and Aphrodite while off to the left side you can see her husband, Hephaistos, at his forge, gesturing angrily toward his unfaithful wife or perhaps casting his numerical net.

SIX

A NET TO CATCH LOVERS

Numerical facts about the planets prove complementary to myth because they have been shown to represent a factual domain to which these myths allude. Thus, ancient numerical and astronomical practices are the core science behind mythology, an extension of the thesis found in *Hamlet's Mill.*

Chapter 4 introduced Bennett's energy system in which the lowest cosmic energies were said to be experienced as conscious energy (understanding) and creative energy (profound action). We now ask why the planets evolved number systems, especially systems based upon the Golden Mean. The creation of the Moon appears to be a particularly strong case of creative energy at work in a profound action, out of which emerges life and the biosphere. If the universe is here to create things, then this act fulfilled that purpose utterly.

Beyond creative energy, Bennett describes something called unitive energy, or "the power of love," which holds the key to how all diverse things in the universe should work together. This integration precedes the creative energy of the solar system and therefore belongs at the level of the galaxy we live in. In myth, this level is Ouranos, who is castrated by Chronos/Saturn. His generative organ is thrown into the sea to become Venus/Aphrodite, who was "born from the foam"—that is, between the galactic sea and solar Earth.

The generative organ is the key characteristic running through this tale of planetary creation—the Golden Mean in one form or another, a Golden Mean that will become the property of beauty in the domain of life on Earth.

Why is the predominant form for galaxies a spiral? Interestingly, the spiral is a common manifesting form of the Golden Mean. Here is evidence that this proportion is the unitive energy of love on Earth, an energy that also belongs to the galactic level of existence where stars are born.

Throughout recent ages, monumentalism has employed the Golden Mean to represent the sacred. This previously seemed arbitrary, but the action of such energy is irresistible, for it is higher than the planetary creative energy. Yet it is strongly manifest in life and, therefore, "love" is all around us. Human cultures have lost the conscious energy needed to understand this.

Ouranos is transformed in myth to become Prometheus, who steals fire from heaven but then has to pay the eternal price of being chained to a rock (the Moon?) to have his liver (the organ ruled by Jupiter) pecked out by day and regenerated by night. Ouranos has thereby become a "Green Man." Life can regenerate itself but it cannot escape its place in the scheme of things. While Earth can be a prison, it is also a place of transformation for human understanding.

The archetypes of man are in motion like the planets. This ancient concept of an inner "planetary system" implies that myth can happily use astronomical allusions in parallel with psychological and socio-historical facts. Mythic characterization, fed by these different levels, alludes to why things happen.

So, the aesthetics of number rule absolutely, with the Golden Mean as the apparent godhead and the planets as its priests. Everything is judged according to harmony, and this is the root of the prehistoric value system, ethical and cosmic. Additionally, the cosmos continually seeks creative actions to produce new harmonious outcomes.

This is why the Argonauts had to seek a golden fleece and why Hephaistos had to marry Aphrodite. But how did she go on to tame the unsubtle and wild Mars and then get caught doing so in a net made by Hephaistos?

CATCHING AN ACT OF LOVE

The golden throne establishes harmony with the pentagram of five Venus synods, which take eight practical years, a cycle that recapitulates the five periods of seventy-three days in a practical year and eight periods in a single Venus synod. Seventy-three days contain four of the 0.618 lunation units created by Hephaistos in his forge. Therefore, there are twenty of these units in a practical year and thirty-two in a Venus synod.

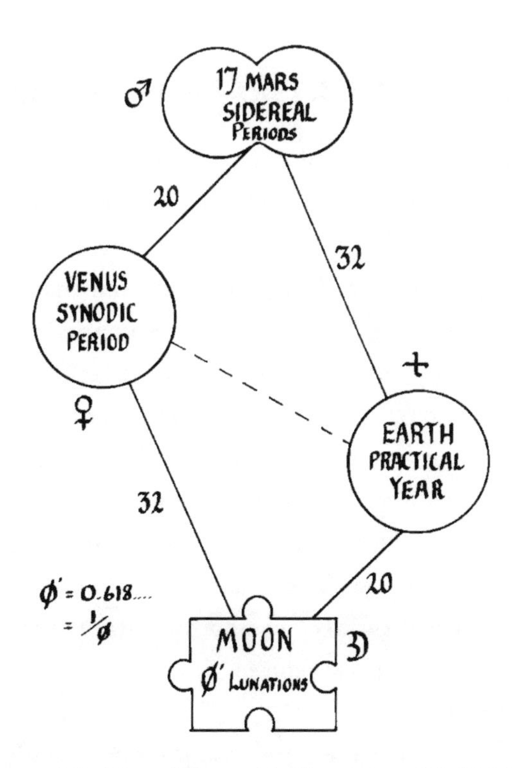

Figure 6.2. Mars Flirts with Venus in a 20:32 Net (Drawn by Robin Heath)

These numbers, twenty and thirty-two, reincarnate in a higher register using another diamond to create a period thirty-two practical years long. This is twenty Venus synods long as well as seventeen Mars sidereal periods! A relationship with Mars is developing here—a sidereal flirtation that could lead to whole number, interplanetary sex. But how can this occur?

These diamond shapes are turning into nets, which is exactly how ancient mathematicians viewed the parallel development of two numbers to form a symmetrical shape such as a square or rectangular number.

The first net of 20:32 catches a lusty fish, Mars. But what kind of net will properly catch Venus? The ratio of a Mars synod to a Venus synod is four to three—it is that simple! In four Venus synods, there are three Mars synods. Having made this net, our lovers are naturally and eternally going to be caught in it—they have made their bed and they must lie in it!

LOVE ON A GRANDER SCALE

Mars does not base his being on the Golden Mean lunation that subdivides the Venus synod and practical year, but appears to hail from the transformed Golden Year itself. We know that the solar hero figure is associated with the number thirty-three and find that the Mars synod contains thirty-three units of four over five lunations. Yet the Golden Year contains twenty lunations, which, when scaled into four-fifths lunation units, return us once more to the number twenty-five, since we find there are twenty-five $^4/_5$ lunation units in a Golden Year.

Mars is the son of a god and therefore his ancestry and power emerge as the twenty-five that Jupiter uses to bind the Moon to his synodic period. But the emergence of Mars's power cannot be shown unless we construct the net of twenty-five and thirty-three around the Golden Year and the Mars synod (see figure 6.3).

In twenty-five Mars synods there are thirty-three Golden Years. The significance of this is that the relationship between male and female at the cosmic level is probably linked to these numbers. Since evolution within the biosphere proceeds out of sex, it is well worth attending to the foreplay.

VENUS AND MARS—THREE TO FOUR

Twenty Venus synods create a period of thirty-two practical years while twenty-five Mars synods are thirty-three Golden Years to within eight days or 99.86%. Converting the latter result to practical years creates an apparent bastard of 53.394 practical years until we multiply it by three to get 160.26 years, which is within 99.84% of five times twenty Venus synods. Since Mars is inherently the most erratic visible planet, this still appears to be part of the intended matrix.

Because of the other nets already cast over the inner planets, there are other correspondences operating in parallel, such as ninety-nine

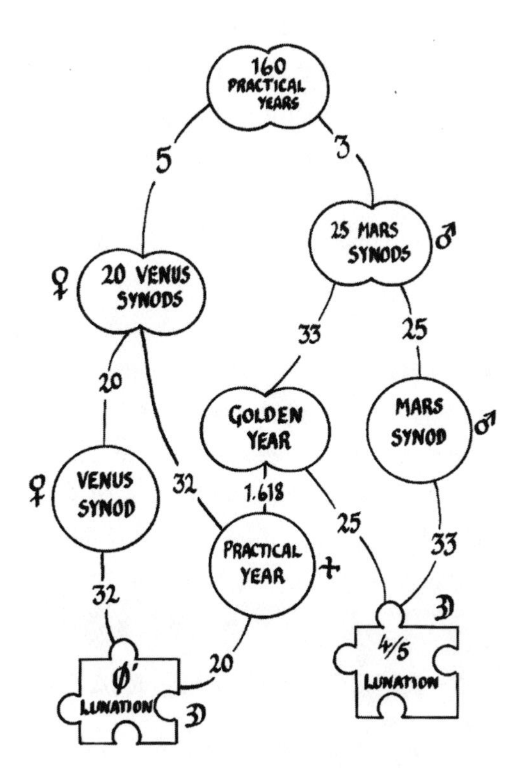

Figure 6.3. The Net That Catches the Lovers (Drawn by Robin Heath)

Golden Years and seventy-five Mars synods. We are in love in an orgy of reconciliation, harmonizing the outer chaos of celestial synodry with a production line of cosmic intercourse. To switch to another world mythology for a moment, Shiva is in continuous coitus with Parvati.

Looking again at figure 6.3, we find some significance in twenty versus twenty-five and thirty-two versus thirty-three. The numerical relationships with thirty-two and thirty-three hold the Golden Mean between them while the numerical relationships with twenty and twenty-five can hold the 3:4 ratio between Mars and Venus. Explicitly: Five times thirty-two practical years equals 160 practical years and three times thirty-three Golden Years equals ninety-nine Golden Years and 160 divided by 99 equals 1.61818 (99.88% of the Golden Mean). Three times twenty-five Mars synods equals seventy-five Mars synods; five times twenty Venus synods equals one hundred Venus synods, giving a ratio of 3:4 between Mars and Venus in units of twenty-five.

While the 0.618 lunation period is common to the Venus synod and the practical year, the new unit of four over five lunations emerges to relate the Golden Year to the synodic period of Mars. The Golden Year as twenty lunations naturally divides by $^4/_5$ lunation to yield twenty-five units. More surprising is the fact that there are thirty-three of these units in a Mars synodic period.

These number systems run throughout our evolution. The lunar year of twelve lunations relative to the number of lunations in a solar year (12.368) forms a ratio of 32.618 to 33.618. We have only scratched the surface, yet we have sketched out the arc of time within which planetary creation took place. Plus we've solved the apparent disharmony of the Moon to the solar year through its Golden Mean relationships to the practical year.

THE ROOTS OF THE MATRIX

In the last chapter, Hephaistos created a 4:5 resonance between the lunar orbital period divided by twenty-five and then twenty. The result is a twenty-lunation cycle within the Golden Year that calibrates Venus and the practical year with 0.618 lunation unit (figure 5.3).

The basic relationship of Venus to Mars is 4:3 but Mars also has a relationship to the Golden Year, 33:25. A common root unit that complements the 0.618 lunation is a lunation divided by five (LUN/5). This unit is present as follows: There are 3 x 33 = 99 of these units in a Venus synod and 4 x 33 = 132 of them in a Mars synod. This numerical unfolding is illustrated in figure 6.4.

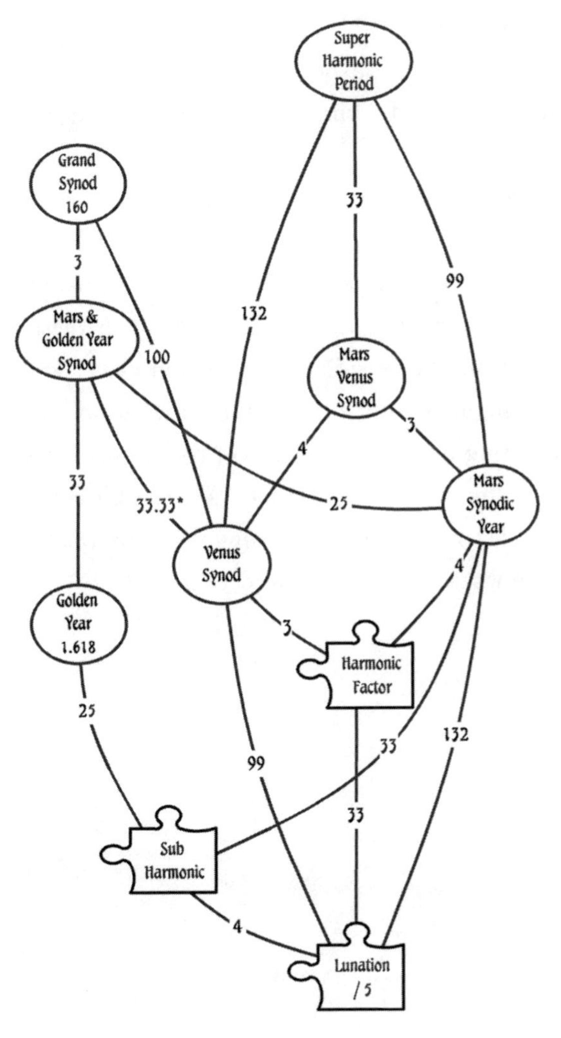

Figure 6.4 The Five-Basis of Hephaistos

Venus divides the zodiac into five parts over eight years. A circle or orbit divided by five, a pentagon, is an extremely significant and enduring symbol—an occult signature that may go back to the first smith Hephaistos, for "smith and shaman come from the same nest."

If we remember that Saturn "cast" Venus (of the pentagon), then perhaps Hephaistos is one of the remnants of Saturn who now lives in a golden cave in Ogygia. The fall of Phaeton is directly relevant here, as Phaeton/Satan, like Hephaistos, fell into water, a river, having been shot down by Zeus for bad driving along the celestial road. The pentagon is the symbol of Satan's work. This all becomes manifest in the later creation by Zeus of the lunation period divided by five. Accordingly, we could now create a calendar that includes all the inner planets except Mercury.

Hephaistos is named Vulcan in Roman myths. *Volcanism* is named after him, a word that still alludes to underground forges. Without volcanism, plate tectonics, and carbon dioxide emissions, there could not be life on Earth. The planet Mars died for want of these very aspects. Life itself lives on the other side of the carbon cycle, with chlorophyll absorbing it through sunlight to feed the entire food chain and simultaneously regulating atmospheric carbon to balance greenhouse gases and hence the climate of our planet. We are currently testing this sophisticated cycle to its limits by burning through much of Earth's fossil fuel supply in less than two hundred years.

SEVEN

THE MEASURE OF EARTH

The second time Hephaistos is thrown out of Olympus, he lands not in water but on Earth. He was a bit lame before, but now he finds himself a cripple. His landing on Earth points to his final role as the Roman god Vulcan, Earth's original volcanologist.

Ask a child who Vulcan is and he will tell you that Mr. Spock is one. The *Star Trek* storyline belongs to modern myth making. The name and behavior of Captain Kirk is a characterization of Jupiter. Uhura corresponds with Athena, the virgin moon goddess. Surely this rates as mythological authenticity, but where does it come from?

Some conspiracy theorists might imagine an elite corps of individuals who intentionally control or evolve the human population of Earth, but this may not be a necessary explanation. Conscious energy resides in what the Swiss psychiatrist Carl Jung (1875–1961) called the collective unconscious mind. This is a mind in that it thinks through the noosphere, a sphere of mental energy around the biosphere that contains the biosphere's cognitive energies. Myth resides in the collective unconscious.

In the second *Star Trek* movie, Spock is reborn on a new Earth by the "genesis effect" with a confusing case of memory loss. But the new

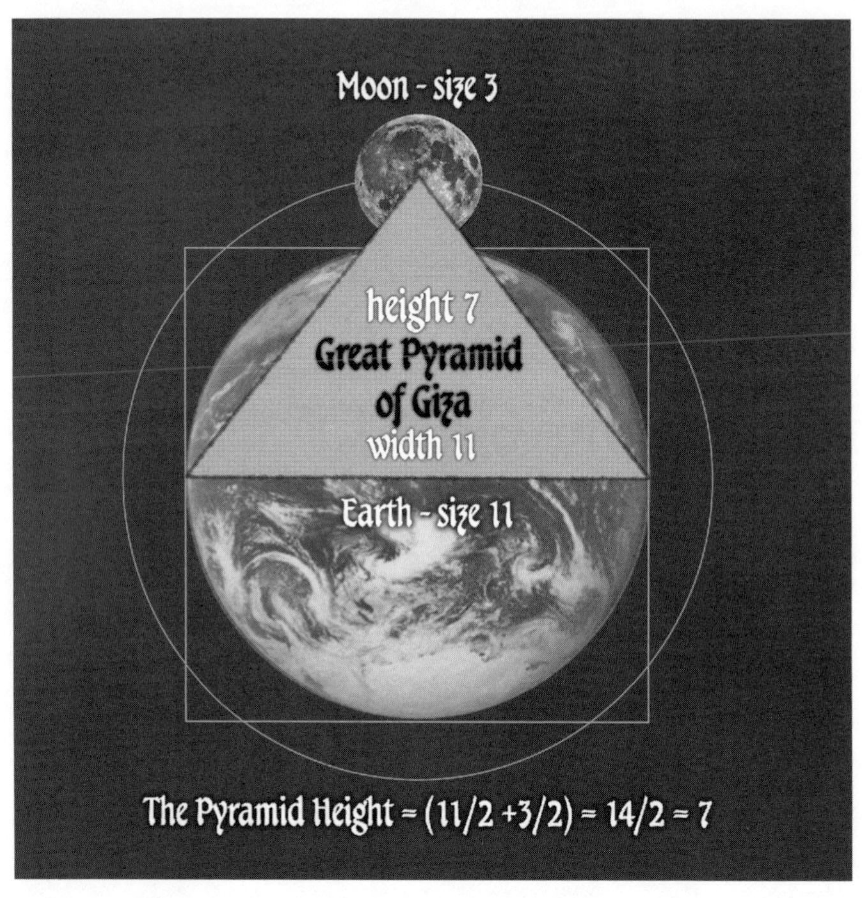

Figure 7.1. The Measure of Earth and the Moon
There is a tradition that says the ancients encoded their knowledge of the world in the dimensions of their sacred monuments. Here the dimensions of Earth and the Moon are shown as a geometric construction of a squared circle integrated with the Great Pyramid's proportions. (After John Michell)

planet shows distinct signs of volcanism, and a race to rescue Spock before his problems destroy him ensues. So how does Vulcan's part in the planetary matrix fall to Earth? Just as units of time in the sky define a planetary matrix, we find that units of length are evolved from the sky and also the size of Earth using ancient measures—bringing more work for Hephaistos and the Golden Mean.

THE SIZE OF EARTH

A culture that wants to measure heaven and Earth faces a problem: There is no real connection between the two. On Earth, an object has length. In the sky, everything is an angle. During the Dark Ages, this fact was given as a reason why what was up there could not possibly be related to what was down here—a doctrine that suited the Church in its role as intermediary.

The only part of Earth that clearly relates to time is the equator. In the course of a sidereal day, relative to the stars, Earth rotates exactly once. While the planets, Sun, and stars come and go as Earth turns, their motion equates to the distance traveled by a point on Earth's surface. This means that angles can become distances that represent time on Earth.

At the equator, a point travels 43,800,000 yards in one sidereal day, if we can accept that the Imperial yard has reached us with an accuracy of 99.94% of its original value. If the ancient yard was an exact subdivision of an ancient measurement of Earth's circumference, then this small level of error is quite possible in either the measurement or the ancient unit.

Now, 43,800,000 yards divided by 365 is the distance a point on the equator moves during one chronon, which is 120,000 yards, twelve times 10,000, or 360,000 feet. Previous chapters have shown that the numbers 365 and 10,000 are ubiquitous in the system of time related to Jupiter and the Moon. The number twelve is of further interest because there are twelve inches in a foot. Using this system of measures, the average single degree along Earth's circumference equals 365,000 feet, connecting angular measure to length (see figure 7.2).

Dividing the 120,000-yard-per-chronon distance by 10,000 and then by twelve creates the measurement we call a yard. Obviously, any planet's equator can be divided in this way. However, in the case of Earth, where we have 365 days in a practical year, 365 chronons in one rotation, and 10,000 chronons in a sidereal month, the revealed unit is not arbitrary and is in current use as one Imperial yard. So, the circumference of Earth

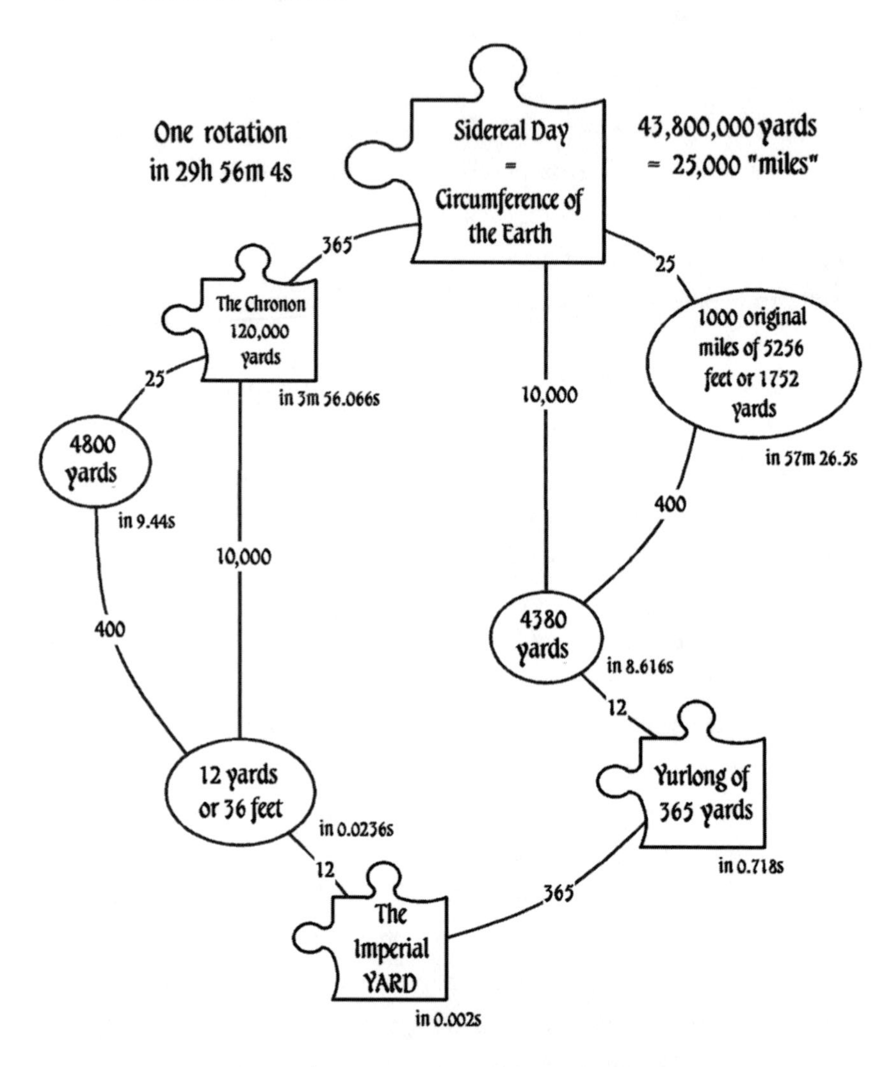

Figure 7.2. The Matrix View of Earth's Dimensions

and the orbital behavior of Earth and the Moon are implicit within the Imperial yard and consequently in every structure built using it. Similarly, the recently outlawed Imperial foot has a fine pedigree based on geometry, the measuring of Earth. The metric system of units has no such affinity with Earth, so Napoleon, who was its apparent champion, referred to it as trivial.

Imperial units can integrate astronomical time relations with physical structures on Earth. In particular, the English "Imperial" yard and its foot are supreme measures that allow access to ancient metrology. As John Neal makes clear in his remarkable book, *All Done with Mirrors,* the oldest surviving standard rod in Britain is the yard, now in the Westgate Museum in Winchester. Dating from before 1000 CE, it is just 0.04 of an inch short of the modern Imperial yard.

Other interesting numbers and units develop from this simple example. The 43,800,000-yard equatorial circumference of Earth is equivalent to roughly 24,886 miles. If we assume the circumference underwent a natural rounding up to 25,000 miles at some time in the past, a mile would then be defined as 5,256 feet (1,752 yards) and not 5,280 feet (1,760 yards). A unit of 365 yards also would surely have been used, further defining the Imperial yard in this example. Since it would be a type of furlong (220 yards), we would appropriately name it the yurlong.

FORGING SPACE FROM TIME

It is important to question the credentials of the Greeks, Egyptians, Babylonians, the Sumerians before them, and of course the ancient Britons. Did these cultures really measure the size of Earth and define systems of weights and measures based on numbers inherited from the prehistoric world? We may answer this question only by investigating how well our legacy of measures fits the known size of Earth. A hefty clue is found in the texts of the first-century Roman scholar and naturalist Pliny the Elder (23–79 AD).

Close to the end of the second volume of his *Natural History,* Pliny writes at length on the distances between cities and the various estimates of Earth's dimensions. He gives 42,000 stades as the length of the polar radius, which is 3949.7142 miles. This figure compares very well with David Clarke's 1886 survey value of 3949.573 miles, and the *Encyclopaedia Britannica*'s quoted value of 3949.921 miles.

Pliny also suggests that the equatorial radius is 3963.5 miles, a figure just a quarter of a mile more than the modern figure. How did Pliny get his hands on such an accurate number? Who made the original measurement and how? Note that David Clarke's survey was undertaken eighteen hundred years after Pliny, before satellites and GPS navigation. He relied on the accurate measurement of angles between an accurate base length together with the sine rule of basic trigonometry, a rule inscribed on Sumerian clay tablets.

Systems of weights and measures—that is, metrology—relate to the work of ancient metalsmiths. But why, around 3000 BCE, did it become so imperative to measure heaven and Earth so precisely?

The people of prehistory used sidereal astronomy, so their work was based upon measurements made against the extremely accurate and stable backdrop of the constellations. In sidereal astronomy, the movement of the Moon is useful for more than simply counting days, and the loops of the outer planets become measurable.

Compare the riches of this science with the way of the last millennium, in which the daytime movement of the Sun was measured from its shadow. This technique is still discussed first in most elementary books on astronomy.

The Sun does not follow a smooth and accurate path because Earth's orbit is not circular but slightly elliptical instead. Also, the Sun's height due south varies through the year. This complicates the measure of time because a seasonal equation is needed to correct any measurements. The obsession with daytime measurements corresponds with the need to divide the day and should no longer be presented as an accurate way to measure time.

Our forebears understood that they lived on a round Earth turning at a fixed rate with respect to the stars. Earth and its spinning mass relative to the fixed positions of the stars describe time in a continuous and highly accurate way. Sidereal astronomy defined time for ancient cultures like the Britons and the Babylonians, and it is the stars that also allow the size of Earth to be reliably measured.

THE TOPOGRAPHY OF SPACE

Sidereal astronomy first defines its asterisms—the unchanging signposts in the night sky. The Moon is seen against a backdrop of ecliptic star groups and these asterisms define its yearly and monthly positions with great precision. They also allow its orbital period to be accurately studied and predicted.

Our asterisms are the twelve constellations of the zodiac. The precursors to the zodiac were systems of twenty-seven- or twenty-eight-star patterns that were recognizable and indicative of a point along the Sun's yearly path (the ecliptic). The Chinese and Indian astronomical systems have retained them, while twelve constellations later became dominant in the West. Also called lunar mansions, these twenty-eight divisions meld perfectly with the division of the practical year into twenty-eight pieces by Saturn's synodic period. Crucially, they also allow the size of Earth to be measured.

In the third century BCE, the Greek mathematician, astronomer, and geographer Eratosthenes (b. 276 BCE) measured the size of Earth using the shadow of the Sun at two places along a line of longitude—that is, at two different latitudes—when the Sun passed overhead at local noon. He noticed there was no shadow at Aswan, Egypt, on the summer solstice. Using the distance from Aswan to another latitude and the shadow angle there, he estimated the circumference of Earth. In so doing, he cleverly avoided the measurement of time, measuring using just one star, the Sun, once a year.

The Sun's position is not a reliable measure of time. Such measurements are more easily made with finer instruments such as those of Babylonian astronomy that observed the stars using long, and therefore accurate, sight lines held at measurable angles to the sky. These can measure the culmination of local noon for any star as it transits the south. With this method, the measuring places can be on the same latitude. Using a fire uncovered between hilltops or monuments, perhaps across a desert, and highly accurate water clocks to measure the time between such culmination events meant that measurements could be made many times in the same evening and not once a year, as with solstice measurements. Stellar astronomy naturally developed the concepts for measurement of both latitude and longitude on Earth. The measurement of longitude requires a line of sight for communication and a clock for accurately measuring short durations. This reveals the size of the equator, which is where time and angle truly meet an equivalent length on the ground.

METROLOGY AND THE MATRIX

The size of Earth was thus measured in prehistory. This led to a number of different units of length, each associated with different ways of looking at the sky and the planetary matrix.

The Egyptian unit called the remen was 14.58 inches long. Rounded to 14.6, it would be equal to 365/25, a major ratio emanating from the Jupiter–Moon relationship. The remen produces a version of the planetary matrix based upon the equatorial circumference of Earth as 108 million remens. The preeminent number of the Moon is 1080 as we saw in chapter 5, and 4320 remen equals 5256 feet—an "old" mile. Both 1080 and 4320 are important canonical numbers in Plato's later numerical systems. Interestingly, a yurlong (365 yards) equals 1080 Grecian feet, each of 1.01376 Imperial feet.

This brings another coincidence to light: The Greek cubit is equal to 18.25 inches—the number of days in 0.618 lunation. This number is related to the time when Hephaistos's mother elevated him back to Olympus and made a finer smithy for him with twenty bellows. Surely, this symbolizes the twenty periods of 0.618 lunation that make up a practical year. Thus, 365 inches equals twenty Grecian cubits, as 365 days equals twenty units of 0.618 lunation, and, therefore, processes can be mapped simultaneously in both days and 0.618 lunation units.

The number of feet in our idealized circumference of Earth is 1.314 times 100 million. Figure 7.3 shows how this number can be reduced to a Grecian cubit. Then 900 remen are found to equal 720 Grecian cubits, 365 Imperial yards, 1095 feet (365 x 3), and 13,140 inches (36 x 365).

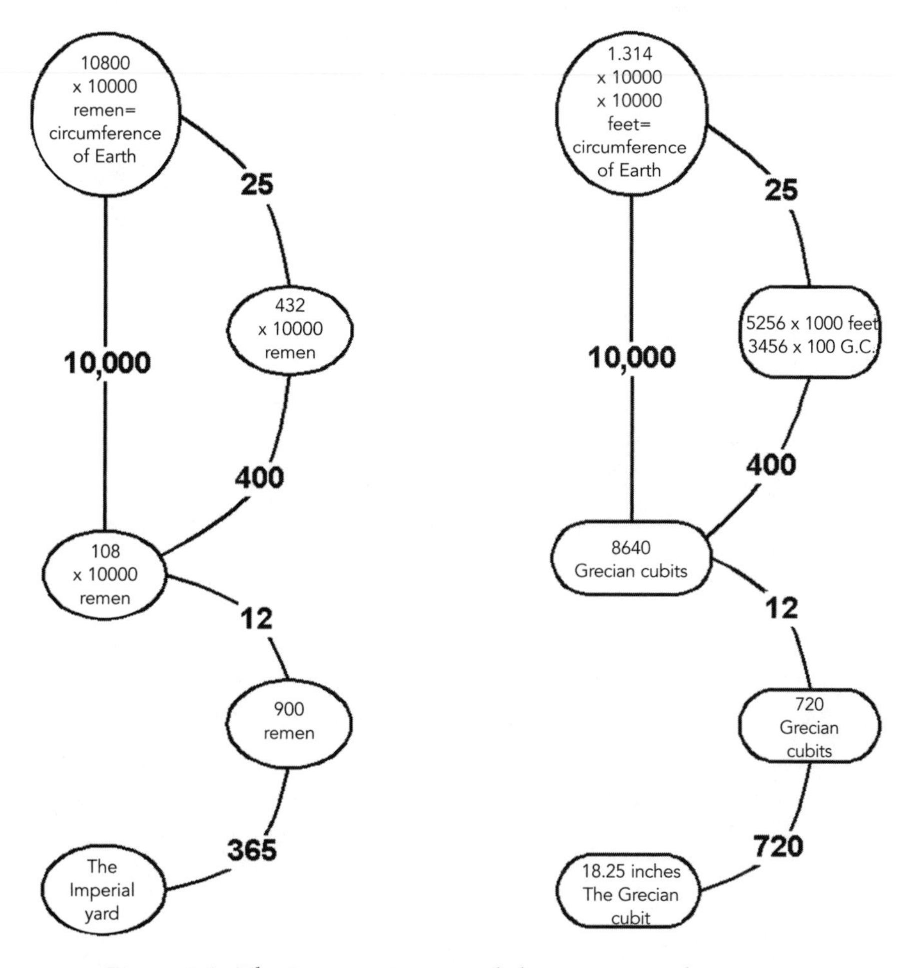

Figure 7.3. The Remen Matrix and the Grecian Cubit Matrix

THE FOOTSTOOL OF THE GODS

All this displays the power of units. They are like great proportional calculators, ancient slide rules for cultures that understood that numerical relationships exist in creation.

If 1.08 times 100 million describes Earth's equator measured in remen, then what unit would make Earth's equator exactly 100 million units long? This requires an understanding of the fractional equivalent of 0.08, or 2/25, and hence the number twenty-five is involved.

Because there are 1.08 times 100 million remen in Earth's equator, a unit with exactly 100 million units would have to be 1.08 times the length of the remen or, as a fraction, 27/25 of it. Since the remen is already taken here to be 365/25 (or 14.6) inches, then the remen should be multiplied by a factor of 27/25 (or 1.08) to similarly define the new unit in inches. This can then be factored in a very interesting way:

$$\left(\frac{365}{25} \times \frac{27}{25}\right) =$$

$$\left(\frac{73}{5} \times \frac{27}{25}\right) =$$

$$\frac{73 \times 3^3}{5^3}$$

This unit is 15.768 inches long, canonically expressed as 365 times 432/10000 inches.

This fits perfectly with the Jupiter matrix and its two 10,000-unit components. We also see that seventy-three is the number of Earth, because otherwise we are left with just the interaction between three and five, both in their highest form, in sacred geometry terms, of the cube or volume.

Demonstrably, ancient peoples accessed the measurements of the sky and Earth to revolutionize the meaning of the world in which they lived. They emerged from a hunter-gatherer, nomadic lifestyle into an agricultural society bonded to Earth. This created more secure economic conditions, the domestication of animals, and the growing of grain, which allowed them spare time during plenty and brought forth a flowering of cultural activity.

THE BUILDING OF STONEHENGE

The complex monument we call Stonehenge already has an endless set of numerical speculations associated with it, but the planetary matrix gives another perspective to this manifestation of divine, celestial knowledge.

The constructional sequence at Stonehenge "imploded" from its initial ditch and bank. The 283-foot-diameter (104 megalithic yards) Aubrey circle was dug around 3100 BCE. Subsequent developments in the area within the Aubrey circle ceased around 1500 BCE. Thus, beyond the mumbo-jumbo about rituals performed by ancient, manipulative priests, as postulated in various archaeological books, lies a discipline of articulation and sequence that must have kept a culture engaged for over 1,500 years.

As we have seen, the matrix of creation began with Saturn's 28/29 relationship with Earth's practical year. This divides the year into twenty-eight sections and gives meaning to the fifty-six holes that form the Aubrey circle. We have also seen how the yearly circulation of the Sun and the counterclockwise movement of Saturn enable the Aubrey circle to simulate the motions of Sun, Moon, and even the lunar nodes. This is documented in detail in my brother Robin Heath's books on Stonehenge.

Three other elements in the monument are also of immediate interest. The famous Sarsen ring of lintel stones (c. 2500 BCE) models the lunation period of 29.53 days using twenty-nine full-sized upright monoliths and a single half-sized specimen to symbolize the lunar month. The inner diameter of the Sarsen ring is fifty-six "long" royal cubits (1.737 feet each) and its outer diameter is sixty of the same units.

Immediately within the Sarsen stones lies the largest ring of Welsh bluestones, which probably contained sixty stones in a circle seventy-five feet in diameter. This ring dates to around 1700 BCE. The number sixty is commonly utilized throughout the ancient world. The Babylonians invented the hour with sixty minutes and the minute with sixty seconds, the ancient Maya independently arrived at a base-60 timing schema, and 360-day years were used in India and Egypt. The Ring of Brodgar in Orkney originally consisted of sixty stones while in Indian sacred buildings we explicitly know sixty as the number of Brihaspati, who is the Indian equivalent of Jupiter.

The division of 365 days by sixty produces six plus one twelfth of a day, or 73/12 in fractional form. Thus, after twelve such periods, we obtain the seventy-three-day period that divides the practical year into five equal sections and the Venus synod into eight. The number five is

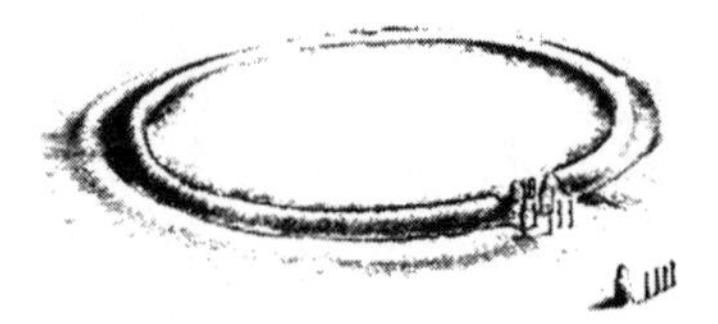

3100 BCE. The ditch and bank of Stonehenge were constructed. The Heelstone (bottom right) defines the midsummer axis of the monument, which appears to be associated with observations of the major standstill moonsets. Postholes in the causeway entrance and elsewhere may have been used to monitor other types of moon setting.

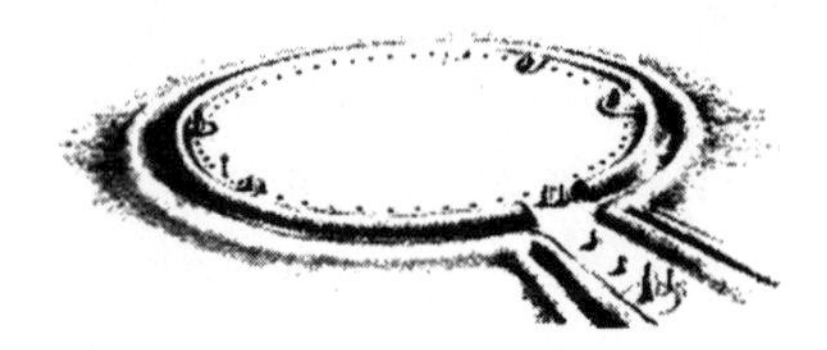

3000 BCE. The fifty-six Aubrey holes were dug, possibly to support a wooden henge. This is 104 megalithic yards in diameter (283 feet). Its geometry enabled the tracking of the motions of the Sun and Moon and the prediction of eclipses. The avenue leading to the circle was constructed at this time. Around 2700 BCE the four station stones were set in a 5:12 rectangle on the perimeter of the Aubrey circle.

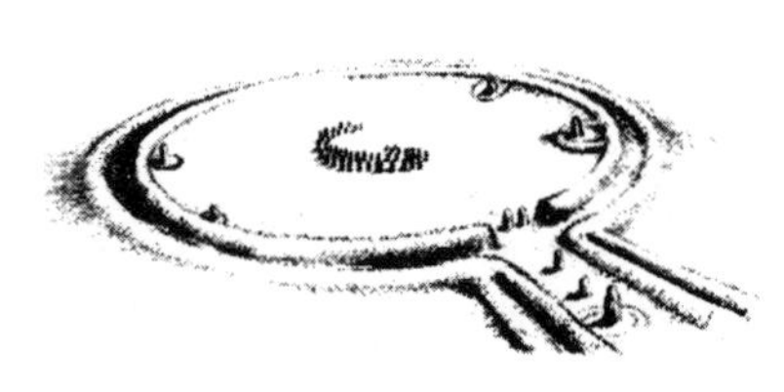

2600 BCE. A bluestone henge was constructed in the center of the circle, possibly replacing a previous wooden structure. Thirty-eight pairs of bluestones were to form a circle ninety feet in diameter. This construction was apparently never completed. Its dating is uncertain because bluestone chips have been found in silt dating far earlier than 2600 BCE.

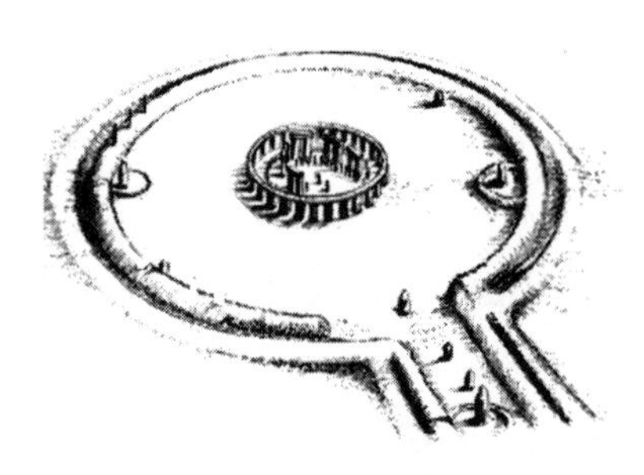

2500–1500 BCE. Around 2500 BCE the Sarsen ring was constructed, and the five massive trilithons were erected in a horseshoe shape aligned to the axis of this 100-foot circle. The bluestones of the earlier henge were recycled into a circle of mainly undressed stones, perhaps sixty in number, and also a horseshoe of nineteen dressed slender stones. These stones can weigh up to four tons.

Fig 7.4. The Evolution of Stonehenge 3100 BCE–1500 BCE (Drawn by Janet Lloyd Davis, courtesy of Bluestone Press)

represented at Stonehenge by the five massive trilithons built in an ellipse just within the bluestone circle.

Furthermore, three of these 73/12-day units combine to create a period of 18.25 days, the 0.618 lunation unit that further subdivides time using the Moon. The numbers two and three, manifest in twelve, alone can enable harmonization with the fivefold year through sixty, which is five times twelve.

Thus a ring of sixty stones, such as at Stonehenge, would have enabled ancient man to represent the new calendar created by Jupiter after Saturn was deposed. This is perfectly and uniquely represented at Stonehenge by rings associated with the numbers fifty-six (2 x 28) and sixty. This numerical material requires very little of the finer geometrical relationships also found at Stonehenge. It can stand alone, making sense of this monument through pure numerical and calendrical circumstance.

By 1500 BCE the building at Stonehenge rapidly came to an end. The project was apparently completed, and no stones were placed after this date.

STONEHENGE AS NARRATIVE

About three hundred years passed after the building of the fifty-six holes of the Aubrey circle (c. 3000 BCE) before the bluestones were fetched from the Preseli Mountains of west Wales. These were initially built into a curious circle of thirty-eight pairs of stones, some topped with lintels. This construction, like the Sarsen circle that followed it, was symmetrically aligned to the midsummer sunrise and midwinter sunset—the earliest evidence of astronomical intent on the site. Four sandstone megaliths, the station stones, were then placed around the perimeter of the Aubrey circle to form a rectangle. These also aligned with the axis of midsummer sunrise. An 1,800-foot banked avenue was also built to further emphasize the solstitial alignment. The builders were clearly experimenting with sky alignments, geodesy, the transportation and use of the so-called lunar bluestones, and a geometrical creation within the station stone rectangle. All of this happened before any clear association of the monument with Jupiter and the Moon.

The station stone rectangle is twelve units by five units of exactly eight megalithic yards each. It also aligns to the extreme setting position of the Moon in its 18.6-year cycle. Stonehenge is sited at a latitude where the angle between the solstitial sunrise and a major standstill of the Moon in winter is a right angle. Not only so, but also the 5:12 ratio of the station stone rectangle measures thirteen units diagonally. This diagonal line aligns to the quarter-day sunrises and sunsets in February, May, August, and November.

My brother Robin has amply shown the many layers of meaning found in this structure, in its latitude, and in the location of the source of the bluestones. But the main point is this: The diagonal of the station stone rectangle, at thirteen units, creates the second triangle in the Pythagorean set. (The first Pythagorean triangle is the 3:4:5 triangle used in primitive surveying to create an easy right angle from a rope of twelve knots or units of length and two pegs placed four units apart and is found in stone circles such as Callanish in the Isle of Lewis, Scotland.) The side of length five in the triangle can be divided into the ratio 3:2, creating an intermediate hypotenuse whose length is 12.369. This is, astonishingly, the number of lunations (full moons) in one solar year accurate to 99.992%.

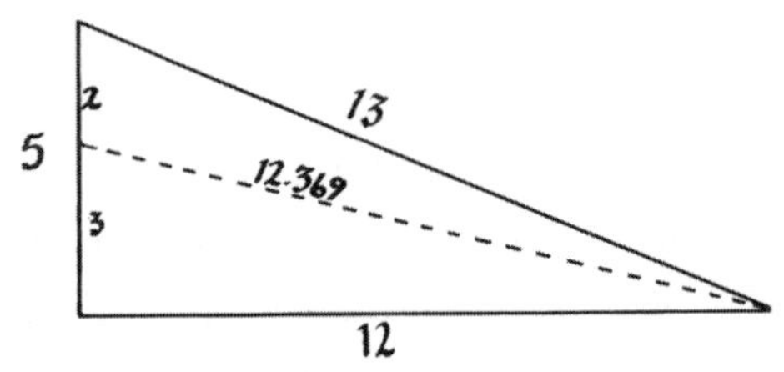

Figure 7.5. The Lunation Triangle

This triangle of four lines is called the Lunation Triangle because the intermediate hypotenuse defined by the 3:2 point of the short side is the length of the solar year if the side of length twelve is taken to represent the length of the lunar year of twelve moon-ths. The intermediate side is, by Pythagoras's theorem, of length $\sqrt{(12)^2 + (3)^2}$ *or* $\sqrt{(144 + 9)}$ *equaling* $\sqrt{153}$ *or 12.36923 versus the 12.3683 lunations in the solar year—a figure accurate to 99.99 percent. Does this length approximate the Moon's behavior or the Moon approximate to it? (Drawn by Robin Heath)*

When this triangle is staked out on the ground with a rope marked in thirty equal lengths (12 + 5 + 13 = 30), the difference between the lunar year of twelve lunations (354.367 days) and the solar year (365.242 days) is obtained from simple geometry and whole numbers. A vital calendrical secret, the difference between lunar and solar years, is revealed. Furthermore, if the rope is marked in units of megalithic yards, the differential length between the twelve-unit side and the 12.368-unit intermediate hypotenuse equals one English foot. Just as 0.368 of a lunation is 10.875 days, so 0.368 of a megalithic yard is a foot. Thus, metrology matches astronomy and time becomes length.

As recorded in the earliest astronomies of India and China, the Moon's motion divides the zodiacal band of stars in twenty-seven or twenty-eight daily motions, or lunar mansions, since her orbit is twenty-seven and one-third days long. This harmonizes well with Saturn's synodic movement. However, more recently, the stars and hence the year have been seen as a zodiac of twelve constellations. Mythology confirms this change: When Zeus deposed the Titans, the twelve gods and goddesses of Olympus reigned supreme and Hercules, the son of Zeus, performed his twelve labors. The myth of Jesus involves exactly twelve disciples, and the myth of King Arthur has twelve primary Knights of the Round Table. In both these cases the leader always forms the thirteenth.

The number twelve is found in the lunar, not the solar, year. There are twelve lunations in a lunar year of 354 days. This leaves a 0.368-lunation overrun or mismatch between the lunar year and the end of the 365-day solar year. Early astronomers would have needed to find the number twelve within the seasonal solar year in order to justify moving to a twelve-sign division of the zodiac and consequently a twelve-month year. The station stone rectangle of Stonehenge broke through to such a new numerical understanding of time, using a megalithic yard of 2.718 feet.

Processes in the material world are based upon a cosmic constant called *e*, which to three decimal places is 2.718. Most commonly encountered in the phenomena of exponential growth and decay, *e* lies at the heart of all cyclical processes. An orbit, for instance, can be modeled as an exponential power where time acts upon two dimensions, one real, the other imaginary. In this model, energy translates smoothly between potential and kinetic forms in two dimensions about a central point. Other oscillating phenomena occur where potential energy accumulates and dissipates, making *e* the fundamental constant of what Bennett called the "hyponomic world of materiality." Similarly, the constant phi, or Golden Mean, lies at the heart of the forms found in the living world.

In the 3:12:12.368 Lunation Triangle, the base of twelve (twelve lunar months) must somehow translate into a twelvefold division of the year. The key to this problem is Jupiter, a key bequeathed through a yard made up of 2.718 feet. When 12.368 megalithic yards are converted to feet (i.e., multiplied by *e*), the result is 33.618 feet, while the superannual "angle" moved by Jupiter in his synodic year is 33.638 days, which is the time Jupiter takes to catch up to the Sun, over and above the solar year.

The derived hypotenuse of 12.368 megalithic yards speaks, in part, of Jupiter as well as the Moon within the year. Furthermore, if we then divide the synodic period of Jupiter, 398.88 days, by twelve, we obtain 33.24 days, thus revealing a relationship to the solar year of roughly 12 to 11. In fact, 11/12 of Jupiter's synodic period is 365.6 days. Jupiter is thus seen to have a period based on twelve units "above" the solar year, just as the Moon has twelve units "below" it in the form of the lunar year. The "above" component, 33.638 days, is related to the "below" component, 10.875 days, by the number 3.09, or five times 0.618 (the fractional part of phi, the Golden Mean) with an accuracy of 99.9%. The lunar overrun period, of 0.368 lunation, is now revealed as the reciprocal of *e* such that one foot (0.368 of a megalithic yard) represents 10.875 days, while the lunation period, 29.53059 days, is represented by the megalithic yard.

There is more reciprocation to discover within the numerical relationships between Jupiter and the Moon. The 10.875-day mismatch between twelve lunar months and the end of a solar year divides into the 365 days of the practical year 33.6 times. Jupiter's synod overruns the solar year by 33.638 days, which divides into the practical year 10.858 times. This strange reciprocation, together with some resulting cosmological equations, connects phi, *e,* the megalithic yard, and key astronomic constants (figure 7.6). If more strangeness is desired, note that the lunar overrun of 10.875 days is four times *e,* and Jupiter's synodic overrun of 33.638 days divided by the 12.368 lunations in a year equals *e.*

The builders of Stonehenge appear to have been studying the Moon from the perspective of the initial circle of fifty-six markers, thereby mapping the motion of both Saturn and the Moon against the stars. Their attention was drawn to the combined solar/lunar behavior found in the lunation. The solar year is literally sandwiched between the twelve and thirteen lunar-month "years" as revealed by dividing the five side of the 5:12:13 Pythagorean triangle into the 3:2 ratio—the ratio of a perfect musical fifth. This gave the ancient astronomers of

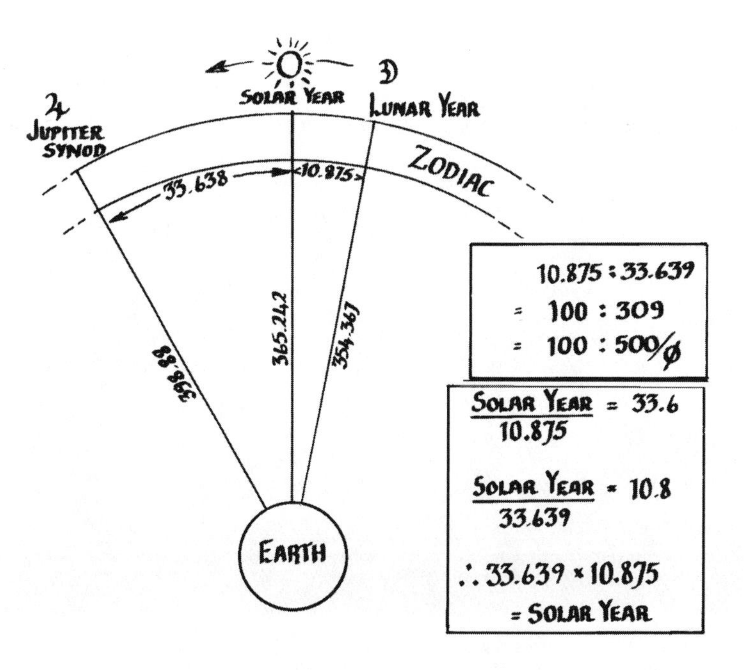

Figure 7.6. The Geometric and Numerical Reciprocation of Jupiter and the Moon around the Solar Year (Drawn by Robin Heath)

Stonehenge an immediate solution to the question "How many lunations are there in a solar year?" However, they had already used the megalithic yard in constructing the Aubrey circle, suggesting a priori knowledge of that unit and the English foot, which gives it its numerical value of 2.718 feet, which allows it to fit within the planetary matrix.

Because the equatorial circumference of Earth has 365,242 feet within every degree of longitude (to 99.999% accuracy), we can therefore surmise that the foot is a more ancient unit than the megalithic yard. The relationship of ancient units of length to the size of Earth suggests that yards, feet, and even inches were derived from astronomical measurements. Obviously, our forebears understood that Earth is a globe. Without recourse to physics, the ancients also found *e* in the sky in at least two places: in the relationship of the lunar overrun period to one lunation and in the relation of lunations in a solar year to the overrun of Jupiter's synod (12.369 lunations to 33.638 days equals the ratio 1:2.72).

Just as the Golden Mean is found sandwiched between the "Fibonacci Venus" of eight over five (or 1.6 practical years) and the twenty-lunation Golden Year, similarly the solar year is caught

Figure 7.7. Outside Stonehenge

Viewed from the ditch, Stonehenge appears as a ruined jumble of gray megaliths that makes little sense to the casual visitor (courtesy Bluestone Press).

Figure 7.8. Inside Stonehenge

Viewed from the central area, Stonehenge appears as a coherent set of concentric circles and ellipsoids, as seen in this view of the Sarsen circle and the axis. Note the Heelstone (visible through the right-hand portal) and two stones of the bluestone circle enclosed within the 100-foot-diameter Sarsen circle. The lintels atop the stones are mortised to the uprights and jointed at their ends. This makes a strong, perfectly level platform fifteen feet "above the reach of the vulgar." John Michell, Ancient Metrology. *(Courtesy Bluestone Press)*

between twelve lunar months "below" and a twelvefold Jupiter synod "above." Both express the constant *e*, or 2.718, which is conveniently incorporated as the number of feet in the megalithic yard employed at Stonehenge and throughout megalithic Europe. The relationship between the megalithic yard and the foot is an analog of the lunation period and the lunar overrun of 0.368 of a lunation prior to the end of the solar year (2.718 can be approximated well as the fraction nineteen over seven and its reciprocal 0.368 can be expressed as seven nineteenths—a fact that gives rise to the important Metonic period of nineteen years in which Moon, sky, and Sun very nearly repeat their orientation).

The station stone rectangle is therefore a transitional object involved in the development of an understanding of time based on the power of Jupiter and the Moon as rulers using the numbers two and three within a twelvefold division of the sky.

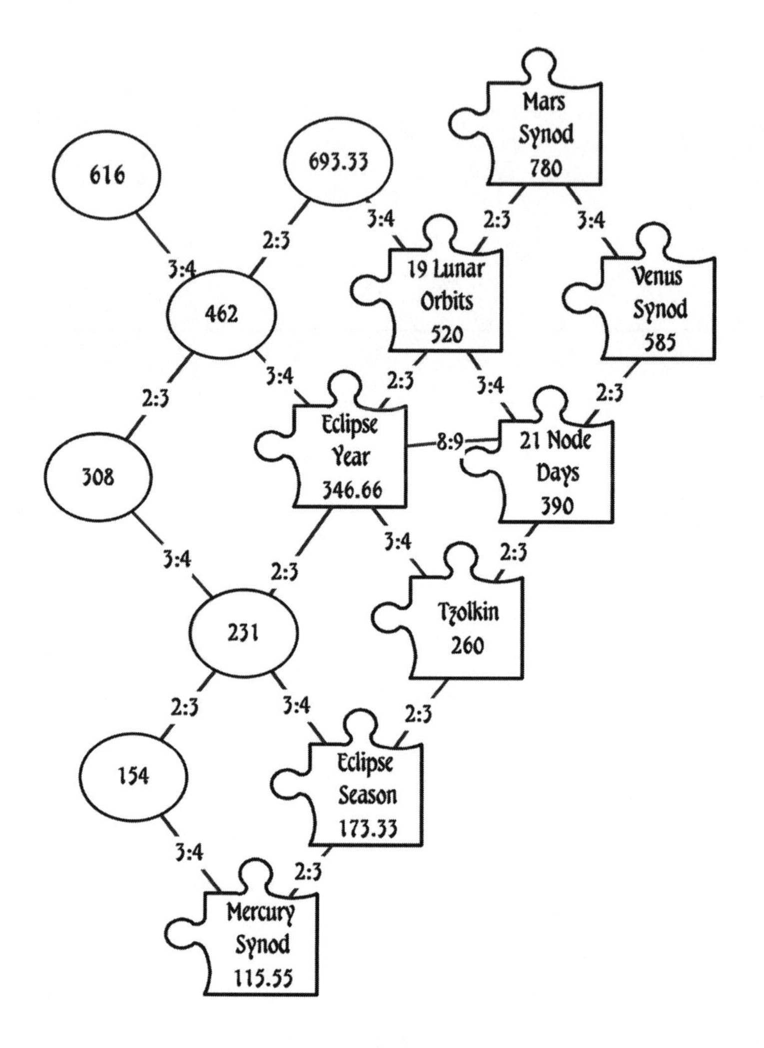

Figure 8.1. The Harmony of the Spheres

The ratios of the musical scale reveal resonance between nature's astronomical constants. The frequencies of the fourth and fifth notes of the scale, fa *and* sol, *form ratios of 3:4 and 2:3, respectively, with the keynote* do. *These provide a matrix that connects the synodic periods of Mercury, Mars, and Venus, the Mayan Tzolkin, the sidereal month, the eclipse year, and the "node day" (the time taken for the lunar nodes to travel one day-degree (1/365.242) around the ecliptic). Note the 8:9 whole tone between the eclipse year and twenty-one node days, with the solar year located at the semitone position (243:256). Items vertically stack as octaves. These harmonic relationships are the key matters discussed within the following two chapters.*

EIGHT

THE GOD OF NUMBER

It is clear from earlier chapters that myths and monuments are filled with significant numerical references that represent a higher intelligence operating through number. Numbers, as cycles, create a planetary world in which a dangerous independence from the source of the creation is possible.

The worldview that the relationship between God and man resembles the parent–child relationship becomes an obstacle in understanding the world as a creation for the angry father-god is only one of the traditional faces of God inherited from the ancient world. There is also the sacrificial child-god, the mother-god, etc., but behind each of these lies a transpersonal Source for the creation in number and harmony.

The world of number enters into all transformational processes within existence while it remains unchanged. In this sense, number is like the metal gold in that it does not chemically react with the physical environment. Similarly, the number seven does not react with other numbers in the planetary matrix and becomes the prototype for the universe as a transformational system on seven levels. It is claimed that the universe was created in seven days and also is made up of reflex states of vibration based on an octave of seven notes, like the familiar musical scale.

Musical theory, yet another branch of ancient science, is an allegory for the planetary matrix because materially generated tones obey numerical "laws" that correspond to the creation of levels of being within the cosmos. Until now, the connection between celestial time cycles and this corpus of ancient musical theory has not been clearly understood because the numbers in the sky were made differently from the numbers in music. For instance, when the year is alluded to in Plato or the Vedas (the primary texts of Hinduism), there are 360 units in a year, not 365 actual days. A numerical abstraction appears to have been introduced, one simpler than that found in the planetary matrix, but when we look deeper, such choices reveal sophistication rather than crudity (see the postscript).

There is a further difference between musical theory and planetary cycles: The vibrations of stretched strings and blown pipes function differently from the vibrations of the celestial cycles. For example, the prime unit of time is the day. It is quite possible for us to say that it will take a week to do a given piece of work such that a week becomes a vibration of seven. But is the week or the day actually vibrating?

Vibrations in a string are dependent upon the string's length, so a frequency such as the day itself is a single drone string that rings on not only through a week, but also in any time period. Thus, a period such as the practical year is receptive to the day's vibration and becomes a period of 365 days.

In the same way, the sidereal day is receptive to 365 chronons. In general, shorter units of time define frequencies within longer time cycles. Unlike a musical string, however, these vibrations are not indigenous to their defining length but instead come from the smaller, whole number time cycles that ring within them.

Two periods in the ratio 3:4, such as Mars and Venus synods, receive vibrations from cosmic frequencies and also give their vibratory ratio to longer cycles. The common cycle over which they repeat is twelve units where one unit can be identified as $^1/_4$ or $^1/_3$ of each Mars or Venus synodic period, respectively. The octave can be seen as from six units to twelve units, a doubling in vibration. In this octave, the ratio of vibrations called a fourth or 3:4 becomes 6:8 (the same ratio with a unit equaling two) or equally 9:12 (the same ratio with a unit of three), according to whether the fourth is at the start of an octave or at the end. Such ratios are called intervals, and they slowly get less harmonious as the unit number employed gets larger. The octave 1:2 is made up of the fifth of 2:3 and the fourth of 3:4. All these numbers divide into twelve, allowing these intervals to divide the octave of six to twelve from above and below to form the two most harmonious notes, the fourth and the fifth. Further to this, the number 360 (as the circle) also divides by twelve, providing another reason to have a year of 360 units.

We see here properties of number where twelve comprises powers of two and three that allow musical ratios to be represented using integers. This concept can be developed to generate alternative versions of the octave. Larger numbers enable more such "fracture lines" beyond the powers of two and three. Employing the next prime number, five, allows an adequate set of musical tones to be built. Ernest McClain, in

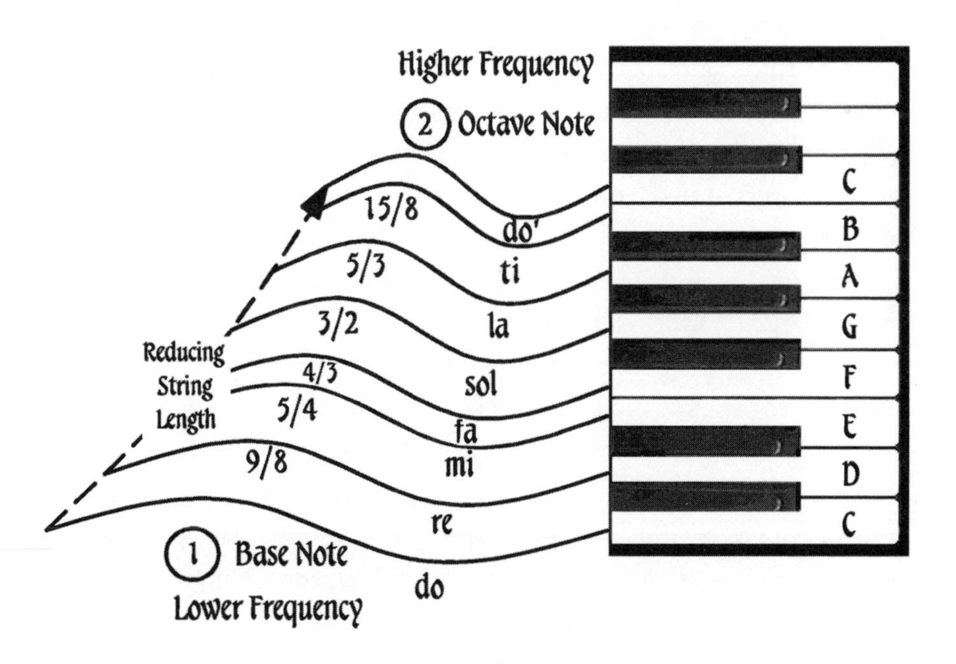

Figure 8.2. Pythagorean whole number tone ratios for the musical scale

The Myth of Invariance and *Pythagorean Plato,* shows that the Vedas and classical works of Plato, Pythagoras, and Socrates hold a fully developed theory of musical tone using integer arithmetic.

Music relies on the ratios between the vibrations of different notes. These intervals involve numbers used to represent the frequencies in simple ratios. Just as the planetary matrix evolved relatively small whole numbers, ancient music theory also has used the smallest possible integers to model musical relationships within the octaves. History has bequeathed us this legacy as coming from the ideas of Pythagoras, although it is likely to be much older than ancient Greece.

THE MATRIX OF MUSICAL HARMONY

"In the creation of the World Soul, Socrates counts 'one, two, three' as do Taoist creationists," says McClain. These three numbers and their multiples provide the key to musical harmony at the primitive level of fourths (3:4), fifths (2:3), and whole tones (8:9) when they are expanded into a matrix containing only powers of two and three.

The fifth is "already there" in two and three, but the development of higher powers is required to illustrate the fourth and indeed the first real octave. As we have seen, the octave is based on doubling and therefore on powers of two, and we may now construct an upward left diagonal from 1 to 2, then to 4, 8, 16, 32, etc. To match this, an upward right diagonal projects the powers of three—that is, 3, 9, 27, 81, and so on. (See figure 8.3 for reference.)

Such a construction belongs to Projective Geometry, a world in which space assumes numerical rules. In this case, moving upward along a left diagonal means "multiply by 2" and moving upward along a right diagonal means "multiply by 3." The ancient teachings, transmitted most notably through the Greeks, are filled with this type of diagram in the form of poetic allusions to what happens within them, just as myth alludes to astronomy.

Two and three are fundamentally incommensurate and lead to the number of the beast, 666, which is a streamlined multiple of two thirds.

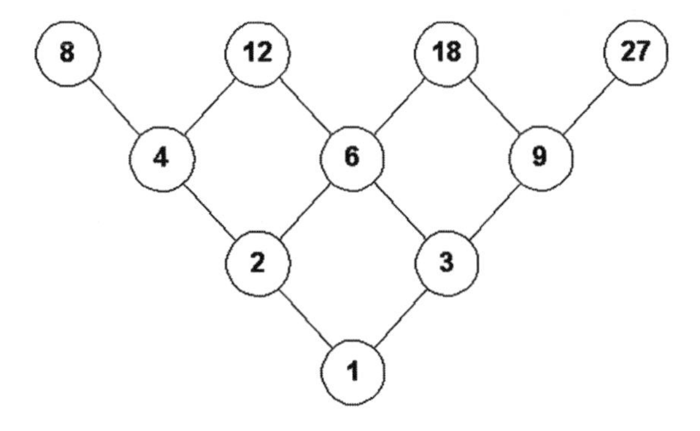

Figure 8.3. The Development of Numerical Musical Ratios

Yet it is this beast that delivers a base for harmony beyond the musical fifth.

Two times three is six, which is the center of the next row in figure 8.3. This forms a new doubling to twelve, an octave (6:12) involving powers of both two and three. Also on row three is four, enabling the archetypal fourth ratio of 3:4, which is then echoed twice on the topmost rows between 6:8 and 9:12. This illustrates a strange property of such diagrams: that the same musical intervals always exist between numbers occupying the same relative position to each other.

Thus, adjacent horizontal numbers represent a musical fifth, while numbers one position to the left and up one left diagonal from each other are two-thirds times two or four-thirds times each other and hence are perfect fourths. The two fourths in the octave 6:12 (6:8 and 9:12) are themselves separated by the ratio 8:9, which is the primordial definition of the whole tone. We can then see how a fifth is made up of a fourth and a whole tone. For example, 6:8 times 8:9 simplifies to 2:3, the ratio of a fifth.

What has come down to us from these ancient works is an obsession with integer arithmetic. From this we find an Aristotelian concept of abstraction from the physical and real world, but, as Gurdjieff explained, the ancient teachings were created deliberately as a legominism—his own word to express the hiding of deeper truths within artifacts that could be reliably transmitted during periods of reduced understanding. These legominisms were said to originate in later Babylonian civilization.

The need for pure integers is in part provided by the science of units of measure since ancient metrology could recover common factors in revealed harmonic relations, such as those in the planetary matrix. The 3:4 ratio (or fourth) found between Venus and Mars, between the practical year and the Io–Europa period, and between others too can be placed onto a similar diagram in place of, say, 6:8 to yield many useful meanings (see figure 8.4).

In figure 8.4, the units are the Venus synod or the practical year, either will do. We notice that every note or period defines a matrix of tones using just the powers of two and three, and within these matrices time becomes significantly divided. Other celestial periods can then relate to points within the matrices and thus form musical harmonies.

All this potential exists in the domain of number and therefore precedes the creation. The number 1 in the center of a relative tone matrix is surrounded by symmetrical numbers and fractions—for example, we find 2 diagonally above left and 1/2 diagonally below right. What we come to recognize is that the "one," so important within expositions of integers in the ancient world, is the base unit with which other units are in musical or other integer relations. This becomes noticeable only when a culture develops a science of scaling or proportion. Thus, when

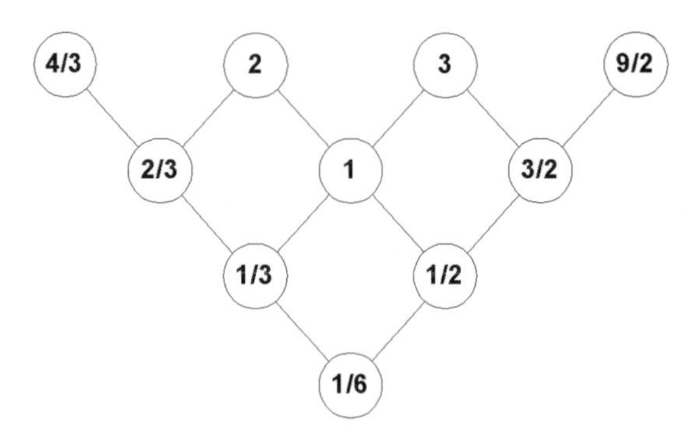

Figure 8.4. Musical Scaling Through Pure Ratio

we are advised to "have a sense of proportion," we have to be able to discover the important factors that divide into something. Once educated in the relativism of musical tone, new facts emerge naturally, such as that 3/2 times a Venus synod is in a whole tone relation (8:9) to a Mars synodic period.

The key legominisms appear to be mythology, metrology (monumentalism based on cosmic units), and musical theory. These become different views of a single science that interprets everything in the created world on a numerical basis. Its origins lie in counting time through celestial events, the discovery of pure number, and the domain of number itself.

The reasoning behind ideas such as the "seven days of creation" is that there must be a process for bringing the universe into existence and that this process parallels the law of just seven notes familiar to music—namely, *do, re, mi, fa, sol, la, ti, do,* which are the harmonic nodes we call notes. Ancient cosmological knowledge asserts that music is just an inevitable audible manifestation of levels of invisible vibration operating throughout all cosmic processes. The planetary system shows these number relations as perfect fourths (3:4 ratios). They organize time through planetary numeric interrelationships and use successive Fibonacci relations that, as we saw in chapter 6, are related to a period of 160 practical years, which is also one hundred Venus synods.

THE RAY OF CREATION

Gurdjieff's sources explained the creation story as God changing the laws governing the universe. Any universe needs a scheme for all its transformations, with differences between vibrations signifying the different states of energy, matter, and information that are possible within it.

Before the creation, before the "In the beginning" of Genesis, the suggestion would be that the universe transformed states using an equally divided octave, with each note one seventh apart. The whole of God's being was present at every point in the form of three forces—affirming, denying, and reconciling—so there was nothing unexpected or new generated in such a system. For this reason it would eventually run down like the spring in a clock. This was a primordial problem to which our universe should be seen as a solution: how to create spontaneous events without intervention.

So the prime law of transformation was changed to the diatonic octave that is more familiar in our world of sound. It uses fractions based upon two, three, and five, rather than seven. With a diatonic octave having note ratios of 8:9 (whole tone), 4:5 (third), 3:4 (fourth), and 2:3 (fifth), relative to the defining base note, the field of number can develop creative "fault lines" where whole numbers can interrelate, through factors of two and three, to create all the primary musical intervals.

The base note *do* doubles with each octave, becoming a new base for a higher octave and the repeating on that level of the same octave structure. The best way to represent an octave is as a circle with the higher and lower *do*s at the same topmost point. But we have already seen that the circle was a representation of astronomical cycles in ancient times and that such cycles divide each other so that a period of seventy-three days, for example, has a relative period of one fifth of a "year circle." This provides the unit for the year-to–Venus synod ratio of 5:8, an interval of a fifth plus a semitone, since (2/3) x (15/16) = 5/8.

Employing twenty-four divisions within the whole zodiac enables the simplest ratios to define the diatonic seven-note scale (see The Musical Key to Numerical Harmony). By expanding this to 360 divisions within the whole zodiac, ancient man could define a circle that measured musical intervals in the heavens because 360 has so many factors and can be described as 2 x 2 x 3 times 2 x 3 x 5. The range of numbers between 360 and 720 then provides the lowest range allowing the generation of a chromatic octave using simple ratios of whole numbers. It also has a unique relationship to 365 and 366 and hence the orbit and rotation of Earth: 360:365 is 72:73 with a unit of five and 360:366 is 60:61 with a unit of six.

An "equal-tempered zodiac" of twelve equally spaced signs relates octaves not to celestial rhythms but to modern musical requirements,

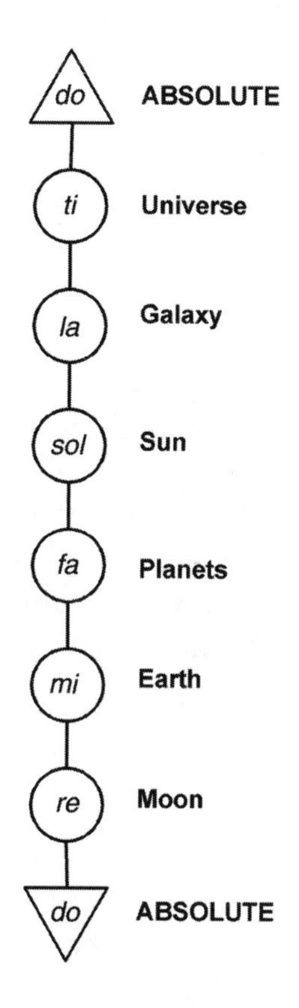

Figure 8.5. Mapping the Cosmos as a Musical Octave (after Gurdjieff)

abandoning the use of integer ratios for an abstract ratio of two to the twelfth root. The equal-tempered system of tuning allows musicians to modulate or change key. However, God did not need to change key because the Absolute is the high and low *do* of the universe (figure 8.5). The simpler, ancient diatonic scheme between fixed points or notes in the creation allows the small numbers 2, 3, 4, 5, 6, and 8 to operate the transformations in the new universe, without 7 or the twelfth root of two.

This simpler scheme and its transformations could all be achieved before the creation of a spatial universe and hence before geometry as we understand it. This achievement is also prior to time and confirms that numbers exist in the "mind of God" and form an Absolute for the creation itself. It is likely, therefore, that modern musical tuning belongs to the modern analytical thinking developed by the Greeks and that, as Plato points out, each tuning system is a "city" in which the citizens live out the consequences of their own numerical orientation, a musical fate set in motion by its "tuning."

LETTING NUMBERS DO THE WORK

The "will of God" in Gurdjieff's system solves the problem of how to make the interrelations between integer numbers "do work." The resulting phenomena that emerge are no longer being driven "from above" but instead drive themselves, with their own will, mind, and so on, all derived from a numeric field "as it was in the beginning."

It is possible to see the creation as the phenomena outputted by a cosmic computer through the inherent properties of number, like the operating system of a modern computer. These phenomena have evolved numerical characteristics that further interact and resonate with other numerically compatible phenomena. In this worldview, everything relies upon other parts of the whole to progress its own nature, leading to an ethic of interrelatedness called the "Law of Reciprocal Maintenance" by Gurdjieff. In this law, all must eat and be eaten on some level, at given times and in different ways.

Numbers were discovered in the sky because the planetary system was already directing these numbers at Earth for the purpose of biospheric evolution under the mandate of a cosmic process generated from pure number.

Earlier we saw that the chosen mechanism for managing numbers is the perpetual motion machine of a planetary system where, after a chaotic emergence, number relations become stable for billions of years, allowing phenomena to develop in a stable numeric environment.

As the numeric results overlay each other, our objective world emerges with a complexity that hides the underlying simplicity that generated it. The sky is really the only way back into the originating subjectivity of the Creator or Mind of God via the work of the planetary demiurge, whose job it was to implement the planetary matrix.

So the planetary world is an involutionary process in which the cosmic descends to the material level by "in-forming" it—that is, determining its form. What goes down can also go up, and indeed this is the true meaning of evolution. The biosphere produces systems that can do work and make higher energies from lower energies with the help of yet higher cosmic forces. At the mechanical level, such energy products are called by Gurdjieff "food for the Moon," since they help maintain the planetary matrix in some way. Specifically, human evolution involves generating energies that relate beyond the level of the planets and the Sun to a selfhood that can approach the Source of all that exists and the creation of the Soul.

While human psychology operates in ways that relate to energies in the matrix, the desires that drive modern human behavior are destructive to both that environment and higher human potential. The system is then in a double bind—man could destroy the environment that created man, but to destroy man would result in the loss of an evolutionary potential. The only solution is for human evolution to break free of exactly those materialistic desires that are currently being amplified through technology. Mankind has made its worldview abstract through analytical thinking and has consequently lost touch with its ancient roots in the sky.

THE MUSICAL KEY TO NUMERICAL HARMONY

The number twenty-four generates the major scale as shown in figure 8.6, in a pure tone matrix that is filled with useful harmonies.

The matrix has two forms of whole tone, with ratios of 8:9 and 9:10, and one type of semitone with a ratio of 15:16 (which is the half-tone of 9:10). The "inner lines" of the matrix are the fifth of 2:3, the fourth of 3:4, the major third of 4:5 (8:9 times 9:10), and the minor third of 5:6 (8:9 times 15:16). It is soon found that all pure tones are products of smaller pure tones, ad infinitum!

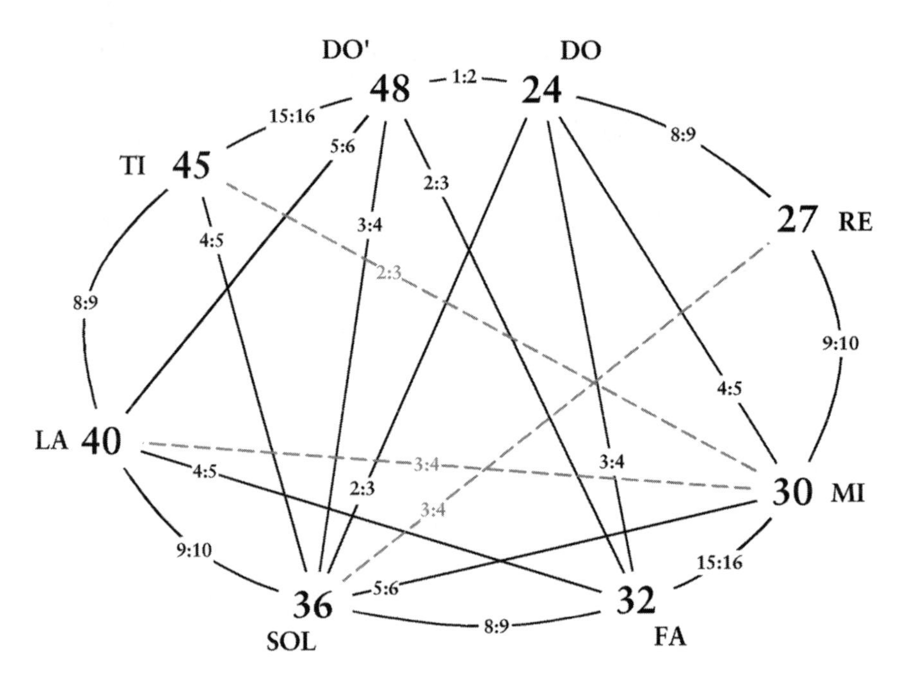

Figure 8.6. The Numeric Matrix of Musical Modes

In modern, equal-tempered instruments, the twelve equal semitones that make up the octave are distributed as 2-2-1-2-2-2-1 to approximate this natural Ionian scale (the Western major scale). If this matrix were based on twelve, then the first full tone could not be generated as a whole number but rather would be 13.5 (see figure 9.5, however, where this does occur with Jupiter). To resolve this and use only whole numbers, we could start with twenty-four to make the Ionian scale (24-27-30-32-36-40-45-48, see figure 8.7). If we start with thirty, in order to form a whole number the semitone must be created first, followed by three whole tones, to produce the Phrygian mode (with a semitone arrangement of 1-2-2-2-1-2-2). Thus, the modes or scales that we've inherited from the Greeks, but which are present in all ancient music, were evolved from an early resolution of these pure tone scales. In this sense, musical scales are created by the lowest whole numbers that can evolve them, and these numbers are then characterized as creative powers that define the existing world.

NUMERICAL ORIGINS OF THE GREEK MODES

Mode	Base	Semitone Arrangement
Ionian	24	2-2-1-2-2-2-1
Dorian	27	2-1-2-2-2-1-2
Phrygian	30	1-2-2-2-1-2-2
Lydian	32	2-2-2-1-2-2-1
Mixolydian	36	2-2-1-2-2-1-2
Aeolian	40	2-1-2-2-1-2-2
Locrain	45	1-2-2-1-2-2-2

Figure 8.7. The Numerical Origins of the Greek Modes
There are seven ways to organize the natural pattern of tones and semitones used in music, starting from whole numbers between 24 and 45 as in figure 8.6. This shows that these modes are properties of the developing field of number itself using relatively small numbers. The rightmost column shows the pattern of semitone arrangement within each modal scale.

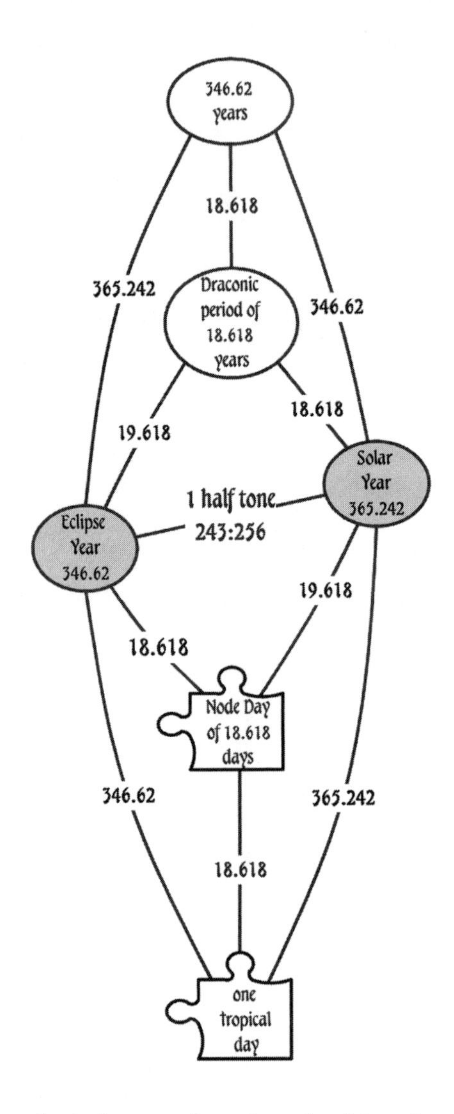

Figure 9.1. It is a little-known fact that the Moon's nodes produce an "eclipse year" (a synod with the Sun) of 346.6 days, which forms a half-tone interval with the solar year (243:256). (These nodes, known as Rahu and Ketu—the dragon's head and tail—in Hindu astrology, are where the Moon's orbit crosses the Sun's path along the ecliptic, causing "the dragon to swallow the sun" if it is behind the Moon.) The number of days in an eclipse year is the square of 18.618. This is remarkable, as 18.618 times 19.618 equals 365.248, which is the solar year. The common period is the Draconic or "dragon" period, the 18.618 years the lunar nodes take to travel once around the ecliptic.

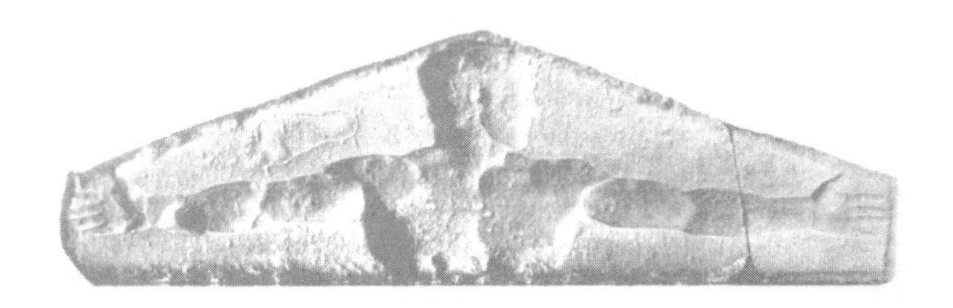

NINE

THE HEART OF THE WHIRLED

As spoken traditions came to be written histories, longer time periods were frequently mentioned with lengths governed by astronomical cycles. These were logical extrapolations of the monthly cycle and the solar year but were of considerable duration and in some cases of such preposterous length that they were clearly mythical in character. For example, the four Yugas or "ages of the world" in Hindu cosmology have a base unit of 432,000 years, which is more likely to be a reference to numerical rather than direct astronomical knowledge.

The "Golden Age" has its origin in the classical Greek, speaking of an original, paradisiacal world age ruled by Saturn, or Chronos. However, references to this age may be traced back to Mesopotamian literature of the fourth millennium BCE, preceding the epic creation story, the *Enuma Elish,* which was written around 2000 BCE.

Another important astronomical cycle, which has persisted in popular culture since the Greeks, is the Platonic cycle or the Great Year. This is the precessional cycle (the time taken for Earth to gyroscopically precess its poles), which causes the autumn and spring equinoxes to move through the zodiac of stars. This cycle in which the stars are shifted in the sky takes nearly 26,000 years to complete. Though its discovery is credited to the Greek mathematician and astronomer Hipparchus (190–120 BCE), there is evidence that Egyptian and Sumerian astronomers knew of it at least two millennia before Hipparchus. The length of the precessional cycle is not absolutely known, nor is it thought to be fixed. It is variously timed at 25,920 years (the precessional Great Year and a convenient canonical number), 25,814 years, or 25,827 years (modern astronomical values). The precessional rotation of the heavens thus takes about seventy-two years to complete a single degree, and inexorably brings new stars to the fore as the Pole Star and equinoctial constellations gradually change.

The 26,000-year cycle of precession can be divided into thirteen periods or "Great Months" of 2,000 years each. For example, the Taurean Age ran roughly from 4000 to 2000 BCE. It corresponds with the period of megalithic building and bull worship. The Age of Aries (2000 BCE–1 BCE) corresponds with the ram, weaponry, warfare, and biblical history. The Age of Pisces (1–2000 CE) aligns with the period of Christianity and its symbol of the fish, which is based on the Vesica Pisces: two overlapping circles, each touching the other's center.

The thirteen 2,000-year periods of the precessional cycle are yet another reason to adopt a thirteen-month year. Our culture, which has adopted a twelve-month calendar, persists in allocating 2,000 years for each of twelve Great Months instead of the true figure, if there were twelve, of 2,167 years. A present-day consequence of this difference is that the celebrated arrival of the Age of Aquarius will not occur for nearly two more centuries.

Hamlet's Mill, the previously mentioned work on the astronomical reinterpretation of myth in history, tells that a primary mythic theme is the precession of the equinoxes and its 26,000-year cycle. The "mill" of the title refers to the ecliptic plane, which is tilted relative to the turning of the sky as Earth rotates, reflecting the tilt of Earth's axis relative to the axis of the solar system. A rotating gristmill often had a top stone angled relative to the base, concentrating the frictional forces in one place to crush the grain.

This metaphor of the gristmill's angled stones is part of a complex of symbols connected within myth. The same image of frictional rotation is found in the fire drill, ancient mankind's most successful technique for making fire. The fire drill is a device that concentrates friction onto a single spot using a stick rotated by a bow similar to a musical bow (figure 9.2).

REAWAKENING THE COSMIC FIRE

Ancient catastrophe myths are centered on fire and flood—universal destruction by conflagration and by inundation. The earliest extant account of this is found in a single sentence spoken to the Greek statesman Solon of Athens (c. fifth–sixth century BCE) on a visit to Egypt. Solon told of a "prodigiously old man" who said, "There have been and there will be many and divers destructions of mankind, of which the greatest are by fire and water, and lesser ones by countless other means." The old priest was drawing on an ancient tradition found also in the Bible and its story of the flood.

Of the four elements known to Greek thought, two are static and two are dynamic. Earth and air are static, with earth representing all solids and air as a transparent medium. Both form the basic stratification of earthly experience. Water, on the other hand, descends while fire ascends, so these two elements become natural symbols of the descent of cosmos into existence and the ascent of consciousness into the cosmic reality.

A cosmology involving mankind would therefore relate to the celestial waters as numerical creations revealed in planetary cycles and to fire as a dangerous but powerful creative energy within numerical systems. The smith Hephaistos made things with fire. The Titan Prometheus was chained to a rock for stealing the secret of fire and thus ending the era of the "golden race" who lived on the "Isle of the Blessed," as described by Plutarch, where Chronos or Saturn still reigned. Presumably, the secret regarding fire was the use of a fire drill, as depicted in figure 9.2, a potent symbol of mankind becoming independent of nature.

Fire was both a sacrament and a sacrifice for the Indo-Europeans

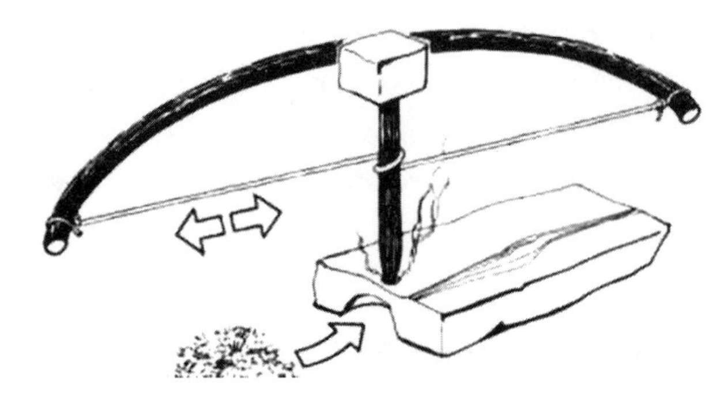

Figure 9.2. A Fire Drill as Illustrated in a Scouting Manual

and hence is common to the Greek, Persian, and Vedic traditions. An eternally burning sacred fire, or one used for festivals marking the year's passage, would be tended by a special social class devoted to the demiurge, such as Brahmins of the Vedic faith. Fire was "thrice kindled," as the flaming orb of the Sun, as the lightning that could set fire to Earth, and as the man-made fire kindled through friction.

FIRE IN THE COSMIC WATERS

If water in the sky is numerical, what is this fire stolen from heaven? The numbers in the sky are based on integers that form the core of the Pythagorean tradition, but, also in that tradition, irrational numbers were considered problematic.

Numbers are called "irrational" when they cannot be expressed as whole numbers, fractions, or a numerical ratio. The irrational numbers most important to the planetary matrix are phi, the Golden Mean, and *e,* the exponential function. These are seen in the golden throne material of chapters 4 and 5 and in the megalithic yard of 2.718 feet in chapter 7.

The words *phi* and *five* even sound like *fire* and, possibly, it is the irrational numbers that are the cosmic fire element complementary to the waters of the planetary world. These "fiery fractions" consume infinite chains of numbers in order to be defined accurately. Irrational numbers form an entrance to a deep complex of ideas relating to rationality itself for, by definition, rationality is the ability to proportion by ratio.

The Golden Mean has already been linked to the Sun and its relationship to the Moon. The Sun is golden, and its associated metal is gold. As part of the galactic level of creation, the Sun acts creatively and is the creative energy discussed briefly in chapter 4. The Sun appears to us as pure fire, burning above the whole number relations of the planetary creation.

By equating irrational numbers with cosmic fire and the Sun in particular, much that is obscure in the different fields of ancient thought becomes accessible, including traditional fears concerning irrational numbers.

The "worst" of the Pythagorean numbers was the square root of two. In musical systems its disharmony can be shown within an octave

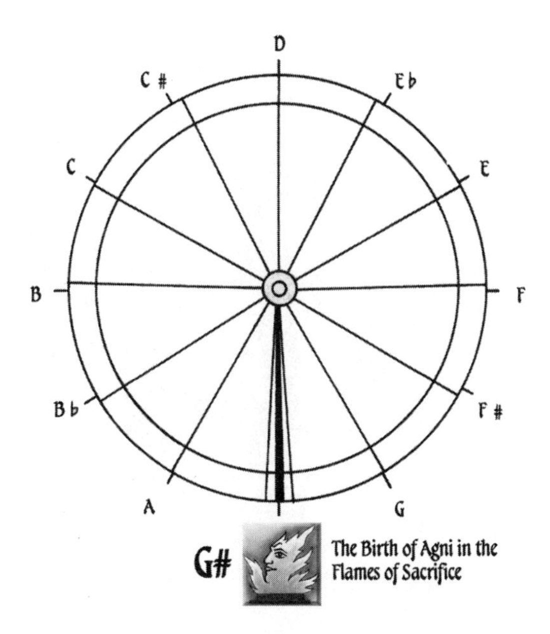

Figure 9.3. Agni, the Vedic fire-god, seen here within the octave, was "born in the flames of sacrifice." Note that the perimeter of such tone circles represents an octave, and the points on the circle are logarithmic angular ratios with the base note at the zenith so that Agni is positioned at the nadir of the circle.

tone circle as the lowest point, between the fourth and the fifth note of the seven-note scale and directly opposite the position of *do* where the octave both starts and ends. The starting frequency and its double at the other end of the octave have a natural geometric mean that simplifies to root two times the starting frequency. (The geometric mean of two numbers is found by multiplying the numbers together and taking the square root of their product.) In Vedic musical theory, this geometric mean note is exactly equated with Agni, the cosmic fire-god.

This complex of fire and water was a primary concept in the ancient world. The cosmic fire drill is seen in the image of the churning of the ocean in the Puranas, the mythical epics of Hinduism. This is illustrated in temple sculpture as a god and a demon holding a cosmic serpent that is looped around a very special stick, just as a bowstring is looped around a fire drill. This stick is supported by a tortoise, which represents the world, and is topped by a four-armed god that may represent the fourfold solar year (figure 9.4).

Figure 9.4. Churning the Cosmic Ocean
In the language of myth, many discoveries within nature and in human culture are attributed to rotational churning, identifying it as a cosmically creative process. Out of it emerged the tree of paradise, the sacred cow, the goddess of wine, the Moon, and in fact every valued thing.

CATCHING THE MOON

An investigation into the relationships among the Sun, the Moon, and Jupiter reveals an extraordinary numerical framework that generates a musical whole tone between these three bodies in a way that would scarcely be expected. This surprising discovery integrates much of what has been described elsewhere in this book and fuses many apparent coincidences into a cosmic design.

It is a great mystery why the frequency of the Jupiter overrun at 10.858 within the solar year almost numerically equates with the period of the lunar overrun of 10.875 days beyond the solar year. The reverse is also true: The frequency of the lunar overruns in the solar year (33.612) almost exactly matches the number of days Jupiter's synod runs over the solar year (33.638 days). This reciprocity can happen only when the product of period and frequency happens to equal the solar year in days. Indeed 10.858 times 33.612 equals 365 days to 99.989% accuracy. Emerging from these curious relationships is a clear implication that the Moon must be resonant with Jupiter. If this resonance could be proved, then much of the mythology about Jupiter might be revealed as coded information concerning cosmic understanding.

These two overruns, when added together, equal the number of days in 1.5 lunations. There is an important reason for this: A Jupiter synod is 1.5 lunations longer than the lunar year of twelve lunations. After each Jupiter synod, the Moon is in the opposite phase as seen from Earth, and the Sun is again, by definition, opposite Jupiter. At that moment, the Moon is in an opposite position relative to both Jupiter and the Sun.

This is symptomatic of the phenomenon called forced resonance, in which a small force is applied in a consistent way at the right time within an orbital cycle to dominate the frequency of an orbit by giving energy or taking away energy, to keep the orbit from changing.

SOUNDING THE WHOLE TONE

The lunar year relates to Jupiter's synod by the ratio 12:13.5 and the division of this ratio by 1.5 lunations reveals a significant fact. There are eight periods of 1.5 lunations in a lunar year and nine periods within a synod of Jupiter. Therefore, this ratio matches that of the classical whole tone (8:9), which is the difference between the perfect fourth of 3:4 and the perfect fifth of 2:3.

This creates a matrix dominated by Jupiter that reveals the inner musical structure of time (see figure 9.5). The octave starts at nine lunations using a base unit of 1.5 lunations. There are six of these units in nine lunations. The lunar year is twelve lunations, or eight base units. So, the lunar year describes an ascending fourth in the octave (6:8).

The Jupiter synod is 13.5 lunations, or nine base units. This describes the ascending fifth. The high *do* note is eighteen lunations, or twelve base units. The tones of the lunar year and the synod of Jupiter form descending counterparts to this note, with ascending fourths becoming descending fifths and vice versa.

The importance of resonance in musical theory is well known. The force to create vibration in a string must be applied near a natural node of that string's frequency or pitch. Such a node is located where a standing wave divides the string by an integer or whole number ratio. In a piano, for example, a point one-seventh down a string's length is the striking point for its hammer. In our planetary example, the separation of the lunar year and Jupiter's synod by a whole tone ensures the transmission of force from Jupiter to the Moon, relative to the solar year.

So where is the solar year in this mechanism—that is: Where is the Sun? The solar year is 10.875 days longer than the lunar year and 33.638 days shorter than Jupiter's synod, totaling 1.5 lunations. It lies between these two periods and very near the "Agni" point relative to the octave, which is root two times the originating *do* frequency. In fact, the number of lunations in the solar year, 12.368, divided by our *do* of nine lunations, equals 1.374, nearly 11/8, and close in musical terms to root two, which is 1.414. However, in natural tuning temperament the

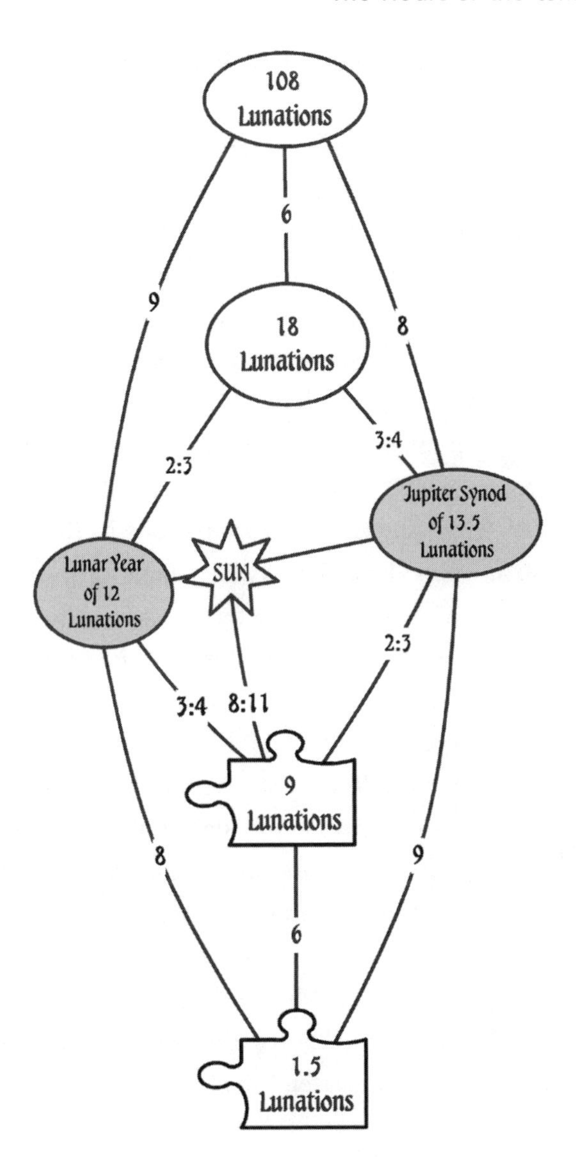

Figure 9.5. Jupiter sounds the whole tone of 8:9 with the Moon. In one of Jupiter's synods there are 13.5 lunations. More precisely, in 136 such synods there are 1,837 lunations. Jupiter has grabbed the Moon to create a musical mill or drill that churns out numerical harmony in an exact, meaningful pattern. This figure also shows the position and importance of the Sun in a way completely different from its place in traditional solar worship. The solar year has been "caught" within the Jupiter–Moon relationship.

relevant note is 45/32 (1.40625), even closer musically to this figure of 1.374 or 44/32.

Therefore, the Sun is in the position of Agni within the octave created by Jupiter's resonance with the Moon. This has enormous implications for the interpretation of every aspect of ancient practice. Just to confirm the veracity of this framework, we note that the top of figure 9.5 is 108 lunations, the perfect canonical number of the Moon found in most ancient traditions. We have already noted that the Golden Mean became known as such because the Sun's motion in one twentieth of a year causes the lunar phases to change by its fractional part—0.618 lunation. We can now see that Jupiter's synod is 1.092 solar years. When this is divided by 13.5 (the number of lunations in Jupiter's synod), it equals 1.618 solar years divided by twenty lunations, or 0.0809 solar year per lunation. The lunation is therefore attuned to the solar year according to the Golden Mean, and gold is the fire of this irrational and unique cosmic constant.

The Sun, in this position (the Agni point of 44:32), is close to one natural semitone above the lunar year and nearly one semitone below the Jupiter synod, demonstrating full harmonic agreement with the celestial waters while also incorporating the cosmic fire. Jupiter and the Moon frame the Sun within a harmonious musical relationship. These two bodies are connected in creation stories and myth with sea beasts (Jupiter) and "the waters" (the Moon). Meanwhile, the Sun, which is the irrational root two, takes the proper role of fire as Agni in Vedic myth. "The fire is within the water," the Vedas inform us.

THE SACRED INSTRUMENT

How does this whole-tone matrix compare with the idealized musical matrices presented in chapter 8? Here, the archetypal musical octave, based on simple integers, uses the octave from six and twelve to develop the first available fourth tone (3:4 as 6:8) and fifth tone (2:3 as 6:9) and hence the first whole tone of 8:9.

However, in Jupiter's "instrument," the number twelve is used as

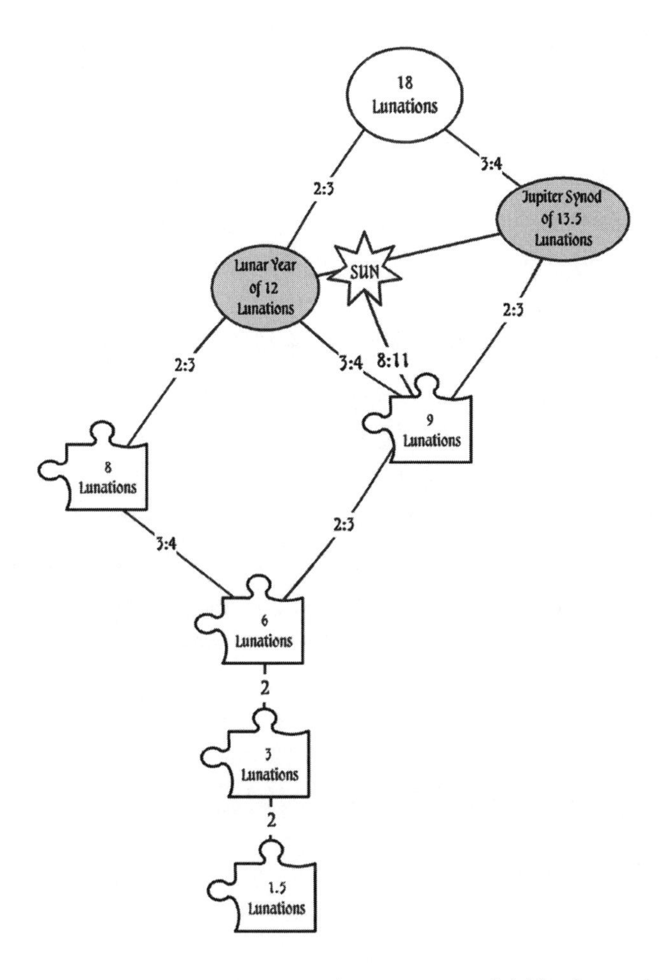

Figure 9.6. Jupiter uses a matrix only one musical fifth above the simplest numerical octave.

the fourth rather than an octave note. Modern musicians consider this easy, since they simply change key or modulate whenever a note in one octave becomes the base note for another octave or key. In the case under consideration, the base note has moved from six to nine, a perfect fifth, making the Jupiter scheme just a fifth above the lowest whole number possibility for an octave, as seen in figure 9.6.

Of course, considering what we now know about the planetary matrix, a fifth is exactly what we might expect from Jupiter, who exudes fives throughout his numerical interactions. But the fifth of 13.5 lunations found here is not a whole number; otherwise it would have been far more evident. Thus, the lunar year of twelve lunations is the only immediate clue to its existence. However, the matrix is immediately adjacent to the perfect octave, one position up to the right as all fifths are, showing its numerical importance despite its inclusion of non-whole numbers.

Another interesting aspect of this matrix is that the octave path rises from 1.5, through three and six, to twelve, so that 1.5 is the root of twelve, the lunar year, which is three octaves above it, and nine is two octaves and a fifth above it.

The absolute simplicity of the lowest whole number octave has been adapted by Jupiter to be the basis of the whole tone that holds the Sun, the Moon, and Earth within his "instrument." The Moon's lunation is the primary vibration within Jupiter's defining synod, with the energy of the Sun and Earth held "in a bow" near the Agni point.

THE COSMIC ENVIRONMENT OF LIFE

Jupiter and the Moon are inseparable because the motion of the Moon is numerically related to the Jupiter synod.

Jupiter takes 361 days or nearly one year to pass through one twelfth of the zodiacal circle, indicating the dominance of Jupiter within a twelve-month calendar. The so-called Sun signs, beloved by newspaper astrology columns, are actually zodiacal signs based on the movement of Jupiter. The motion of Jupiter's synodic overrun divides the year into 10.858 parts. This number, divided by four, closely approximates *e,* or 2.718, so that each quadrant or season of the year equals 2.718 Jupiter overruns.

Because of resonance, these four times *e* Jupiter overruns in one year are related to the twelve plus 1/*e* lunations (12.368) in one solar year. This provides good reason for the megalithic people to define their megalithic yard at 2.72 feet, a measure stubbornly resisted by the archaeological profession because it proves that megalithic people had a metrological culture and that they maintained highly accurate standards of measure over northern and western Europe and beyond. This does not yet fit with a science disinclined to measure monuments with metrology in mind.

Ancient monument building appears to have been driven by the need to find key cosmic facts and to represent them in architecture using measurements, geometry, and location. The ubiquity of the megalithic yard in Britain and the creation of the Lunation Triangle at Stonehenge in particular lead to this conclusion (see again figure 7.5).

The relations alluded to by the megalithic peoples are subtly hidden in the general nature of time. Also, today the Moon is expected to have orbital characteristics that are independent of numerical considerations as she gains energy from Earth's tides. However, in the forced resonance just demonstrated, any such excess energy is taken away by Jupiter in order to maintain the resonant condition. The obscurity of the relationship between Jupiter and the Moon may then correspond with something that happened to time itself: It became occult or hidden information. We shall return to this later. Meanwhile, there is one planet that has yet to be brought into the planetary matrix—Mercury.

MERCURY: GOLDEN MESSENGER OF THE SUN

In chapter 4, we saw that Jupiter is the king of the dark matter of the planetary system, which is ruled by numerical orbital relationships. What of the Sun, therefore, and the kingdom of light matter?

From an earthbound perspective, the inner planets of the solar system, Mercury and Venus, appear to be moons of the Sun. This perception allows us to see their role differently from that of the denizens of darkness—the outer planets—and our Moon, which passes between the light and the darkness every month.

The orbit and synodic year of Venus brings the Fibonacci version of the Golden Mean into the planetary matrix. Venus also introduces the pure archetype of life, the fivefold star, which she traces on the zodiac in five synods. Mercury, however, is more mysterious and erratic.

The path of Mercury's orbit was portrayed in Babylonian culture in the facial features of the "intestinal god," Humbaba (figure 9.7). Of all the planets, Mercury's orbit is nearest to the Sun, even though he appears to dart all around the Sun. For this reason Mercury is incorporated into the characterization of the Mayan fire-god. The wording of the Mayan descriptions reveals that something matrixlike may be lurking here too.

The Maya called their fire-god Hunab Ku. The name means "one palm measuring house," according to William Sullivan, and refers to the hand held out to measure angles in the sky, a symbol of the astronomer-priest. The angular distance from the Sun to Mercury can reach about one palm or ten degrees, viewed from Earth. But Mercury is also said to contain a "measuring house."

Efforts to integrate Mercury into the matrix of creation using most units fail, but the newly discovered unit of 1.5 lunations, which reveals that the Sun is balancing the action of Jupiter and Saturn using the Golden Mean, can be applied to Mercury.

A synod of Mercury lasts 115.88 days. This is the time between sighting Mercury high above the morning Sun on successive occasions. This phenomenon is reminiscent of the burning bush seen by Moses in the desert. Under such viewing conditions, Mercury appears as the flaming sword of the fire-god, full of gold. The ratio between the length of Mercury's synod and 1.5 lunations is 2.616, which is extremely close to the Golden Mean squared (99.92%).

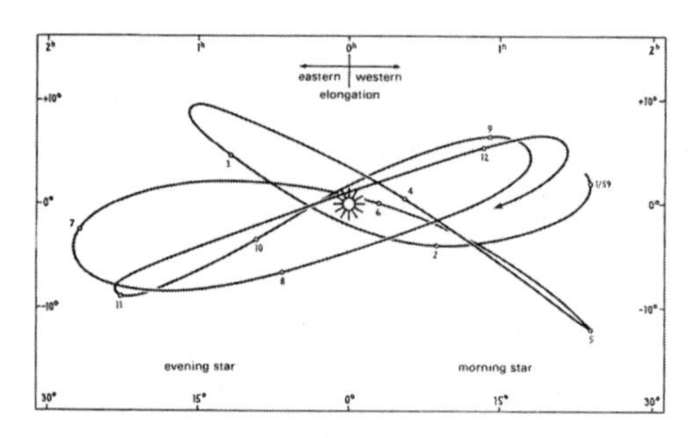

Figure 9.7. Mercury's orbit is very irregular. This is represented in the intestine-like facial features in this depiction of Humbaba, the Lord of the Cedars from the Babylonian epic Gilgamesh.

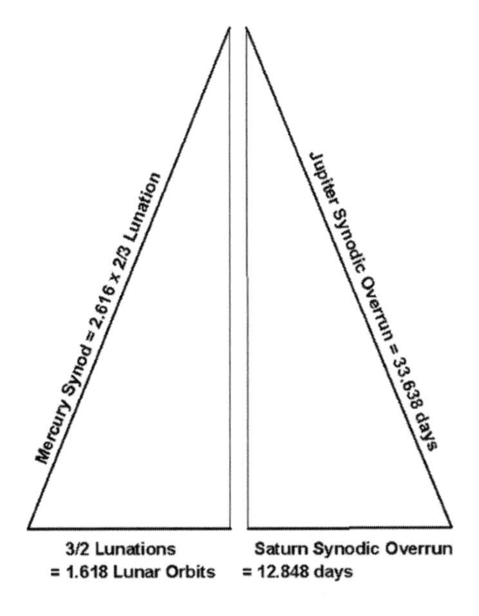

Figure 9.8. Golden Mean Relationships between the Inner and Outer Solar System

This ratio forms a triangle that repeats the relationship between the overruns of Saturn and Jupiter that "passed the measure" from Saturn to Jupiter as seen in chapter 2. The two triangles shown in figure 9.8 indicate that the light and dark forces of the solar system are balanced and also based upon the Golden Mean. Since 1.5 lunations lie at the root of Jupiter's whole tone instrument, with the Sun's orbit on its centerline, it is extraordinary to find this archetype repeated.

Note also that 3/2 lunations is equal to 1.622 lunar orbits because 2/3 lunar orbits equal 0.618 lunation, as shown earlier. The Golden Mean turns up everywhere we look as the cosmic fire. For more Mercurial relationships, see again figure 8.1.

THE FALL OF SATURN

It has always puzzled scholars that Saturn had his own calendar in folk traditions. We are familiar with the seven-day week, but in some ancient calendars, the month was twenty-eight days long and the year consisted of thirteen months and hence 364 days. It was said that a king would rule for "a year and a day"—that is, 365 days, a practical year—during which he would be a consort to the Goddess and a guarantor of fertility before being sacrificed, dismembered, and scattered on Earth. This myth is common to most ancient civilizations.

The story of Jupiter's overthrow of Saturn is similarly found in the myths and legends of many civilizations, but little is told of the consequences this coup had for mankind and Earth. It is possible that the overthrow of Saturn was more than an idea and that it was a historical event. The numerical frameworks discussed in earlier chapters are actualizations of number as planetary orbital periods. This suggests that the nature of time itself could have changed, with enormous implications for mankind.

The thirteen-month, 364-day calendar has regularly been proposed in modern times as more logical than the irrational Gregorian calendar. This happened most recently in 1949. Supported by the British Astronomer Royal, the issue was narrowly postponed by the UN General Assembly. Since then it has withered from serious public concern.

The advantage of the 364-day calendar is that because it is divisible by seven, the whole calendar's structure is rational and integrated (see box on page 25). The major problem with this, which is shared with all other calendars, is that the lunar month is 29.5 days long and not the twenty-eight days required to make it work with the ubiquitous seven-day week. Saturn's synod of 378 days corresponds to 13.5 lunations only if the lunation period is twenty-eight days long. A twenty-eight-day period for the lunar month therefore reveals the sensibility of the ancient thirteen-month calendar in which Saturn, not Jupiter, had once caught the Moon.

The Saturnian calendar in surviving traditions and the extreme monumentalism found post–3000 BCE point to an extremely simple but shocking possibility. The absence of evidence for high cultures before five thousand years ago could be the result of the obliteration of an entire cultural fabric by a natural disaster that resulted from Jupiter's overthrow of Saturn to gain control of the Moon.

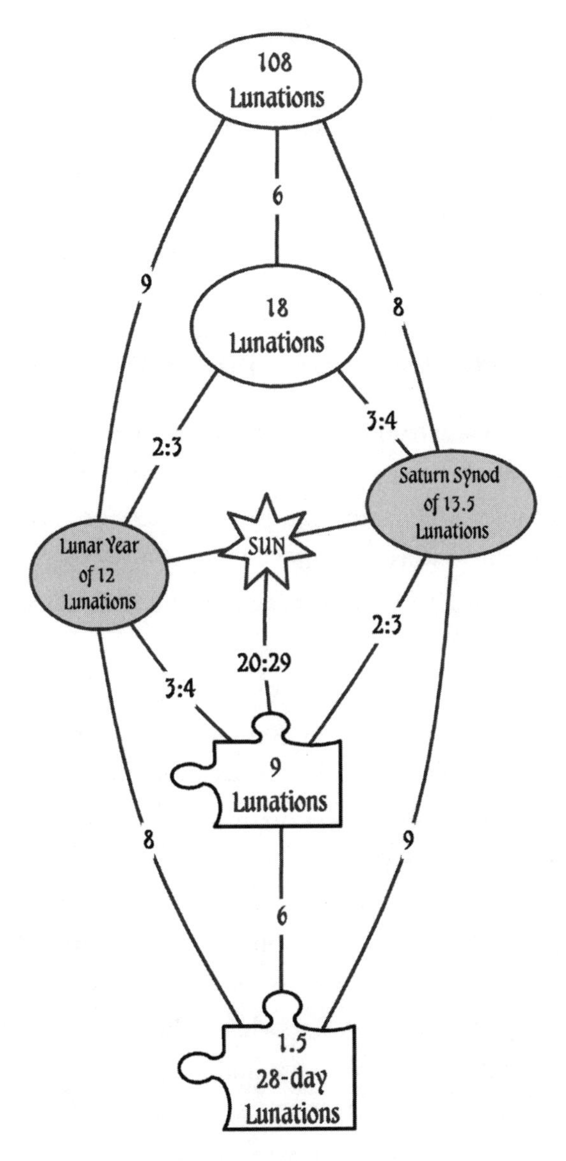

Figure 9.9. Saturn Sounded the Whole Tone until Jupiter Stole the Mill

If such a case were shown to be true, modern astronomers would have much more interest in the building of Stonehenge. The megalithic yard embodied in the architecture of Stonehenge is the key constant emanating from Jupiter's orbit and is echoed in the lunation behavior of the Moon.

Stonehenge's evolution began with a perimeter circle of fifty-six markers (28 x 2), within which the Lunation Triangle is set. The later-built Sarsen circle indicates the change in the length of the lunar month to 29.5 days. The thirteen-unit side of the Lunation Triangle becomes as important as the twelve-unit side, since a thirteen-month lunar year would have calibrated the original Saturnian calendar.

THE WAR AGAINST HISTORY

Why were so many astronomical monuments built and why was there such a flowering of interest in numerical solutions to astronomical and geomantic problems? Imagine the problems created by a catastrophe during which the nature of time changed from being regular and explicit to appearing to be very irregular and having the implicit, and even secret, structure that it has today. The interim period, when the Moon fell between two masters, would correspond to the catastrophe.

The ancients certainly understood the problem of the nature of time and they left their solutions within their monuments. The story of the catastrophe was written into the myths and legends of flood, fire, and the battle between Chronos and Zeus.

Catastrophic destruction by water and by fire need not be separated. Indeed, catastrophism describing both inundation and conflagration reached the ancient Greeks from the same source, Egypt. Ovid relates the story of Phaeton, who, while riding his solar chariot across the sky, was abducted by Aphrodite and appointed nocturnal guardian of her shrine, after which he lost control of his vehicle and crashed into Earth, igniting vast swaths of land and causing widespread destruction. Is this an echo of Zeus's thunderbolts and the punishment of Prometheus for stealing fire from the gods?

In his rigorous book *The Great Year,* Nicholas Campion reminds us, "We can never be sure that some archaic Paleolithic folk-memory is not preserved in mythical form . . ." I have presented a numerical basis for the catastrophe myths. The stories of the Golden Age of Saturn, its golden race of people, and its destruction during the passage of power to Jupiter/Zeus are underpinned by solid astronomical foundations and numerical scaffolding, clearly seen in the ancient monuments.

How would the human mind be affected by such a catastrophe and the subsequent changes to the nature of time? While the old calendar has gone underground, the myths have become blurred, and the monuments are in ruins, our relationship to the world has been transformed since the days of classical Greece by the development of abstract mental powers. Could these mental changes be part of an accelerating transformation of the human psyche and, if so, was an ancient catastrophe responsible for this acceleration?

Whether or not this proves to be the case, the concept that our planetary system is of accidental origin is certainly wrong. Either an improved scientific understanding of numerical creation will emerge or an entirely new type of science will arise. Our view of the planets can never be the same once we understand the numeric endeavors of ancient Britain, Greece, Egypt, Mesopotamia, central Asia, India, Persia, and elsewhere. In these, we see both many gods and only one God, depending on the level of creation being considered.

The matrix of creation sketched in this book is a cross-section of a much larger pattern. I introduced such facts as a traditional backdrop to both present science and dogmatic religions while showing that ancient cultures worldwide appear to have been based upon them.

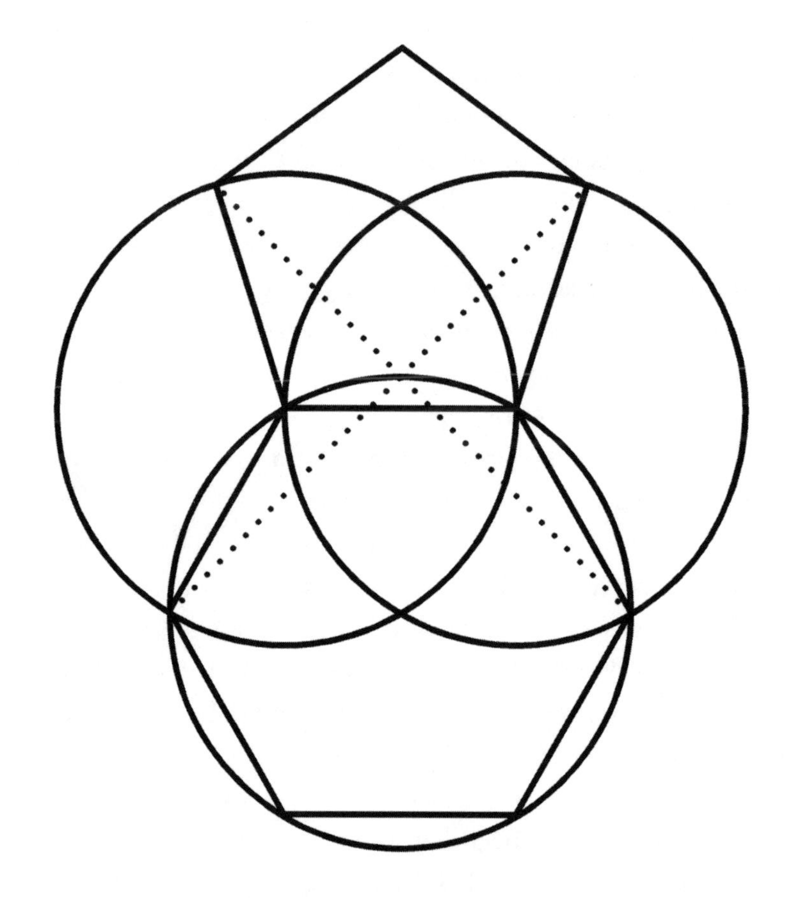

Figure PS.1. The Harmonization of Five and Six

The overlapping of circles, circumference to center, generates the natural sixfold hexagon. The fourfold square, when projected through the circumference, allows a fivefold pentagon to be constructed. Consider also that the practical year can be divided into a pentagon having five periods of seventy-three tropical days, and the sidereal year can be divided into a hexagon having six periods of sixty-one sidereal days. The year was thus naturally seen as based on a perfect 360, having five groups of seventy-two and six groups of sixty. This is probably why the 360-degree circle replaced the 365- or 366-"day" circle. Such harmonizations of number and geometry were signs of cosmic harmony to ancient man and its signature of achievement within the creation.

POSTSCRIPT TO THE SECOND EDITION

The various elements of number, time, myth, and measurement can combine as a single discipline based upon fundamental principles that express the properties of number as the basis for the phenomenal world. For instance, some of the design rules built into Earth's environment are consistent with calendars used by ancient cultures such as those in Egypt, where a calendar of 365 days, the practical year, ran in parallel with another of only 360 days.

These two calendars express a ratio of 72:73 x 5 and form a bridge between the use of degrees and the archaic use of days for angular measurements, especially those on Earth's equator. After seventy-two practical years of 365 days each, seventy-three years of 360 days have passed. The number 360 is more useful since it is a product of the first three prime numbers: two, three, and five (2 x 2 x 2 x 3 x 3 x 5). It also appears to have been the prototype of a culturally defined Earth year. Thus, the question arises, numerically, how many days should there be in a year?

In *The Secret Power of Music,* David Tame suggests that the calendar of 360 days relates to the solar year of 365.242 days as an approximation of the Comma of Pythagoras (the musical interval by which twelve fifths exceed seven octaves, equal to 1.013643265), which is a

The Comma of Pythagoras

The Comma of Pythagoras is a fundamental problem of tuning notes caused by the prime numbers 2 and 3 not dividing into each other.

The unison of an octave employs only the number 2 in the ratio of 1:2. The next great harmony is the fifth, which employs both numbers 2 and 3 in the ratio 2:3.

Prime numbers never divide into each other and yet successive fifths allow a set of notes to be constructed to populate an octave. Twelve fifths span seven octaves, after which the frequency should have risen exactly 2^7 or 128 times the original frequency. However, over the twelve fifths required to populate the seven octaves "fully" and get back to a unison, the frequency has risen $3^{12}/2^{12}$ (129.746), making a difference of $3^{12}/2^{19}$, which is 531441/524288 or 1.013643265.

This problem requires that the notes of such populated octaves be further corrected and it certainly means there is always a conflict between using pure tones and having notes that can be played harmoniously with each other.

In a cosmological sense, it is apparent that the Pythagorean comma expresses a source of disharmony in the world, whose roots can never be adjusted because it arose with the incompatibility of the first two numbers, 2 and 3. Perhaps this also came to be expressed as the number of the Beast, 666. As female (2) and male (3), these numbers are married within their products, notably the duodecimal 12.

fundamental conundrum of musical/harmonic theory (see box). If the solar year of 365.242 days is used then, relative to 360, a ratio of 1.014561 is obtained (within 99.91% of the comma).

In the matrix of creation, however, the practical year of 365 days has cosmic importance, and the ratio of 360:365, or 72:73, is a much more accurate representation of the Pythagorean comma (99.975% of the comma). Thus, the two Egyptian calendars had an implicit ratio related to the Pythagorean comma. Their combined period of 26,645 days times 360 equals 25,902 years, a good approximation of the precessional cycle, and therefore this combined period is equal to the number of years required for one degree of precession.

A NEW MODEL OF EARTH

This is where matters would stop were it not for the work of John Michell and his colleague John Neal. Michell has shown that modern units of measure were derived from the size of Earth's equator. For example, one "angular day" on the equator defined the English foot for ancient peoples. (An angular day is the distance Earth's equator is rotated by in 1/365 of a rotation—i.e., a chronon of time that chapter 7 showed was equated with 120,000 yards.) By their definition, there are 360,000 such feet in an angular day on the equator. Michell also points out that one degree of latitude at Stonehenge and the stone circle of Avebury, Wiltshire (latitude fifty-one degrees north), has the same length as the average latitudinal degree found on the "mean Earth"—the average sphere Earth would be if it were not spinning.

John Neal extended general analysis of historical measures and realized that a system of variations of each ancient foot was employed in ancient times. This enabled him to relate the lengths of different degrees of latitude upon Earth as one moves north from the equator on a grid between certain key latitudes. There is a maximum variation across this grid of 3168/3125 or 1.01376, which is within 99.989% of the Comma of Pythagoras.

This grid variation exceeds any variation in degrees of latitude, so explanation for its use is something of an open question. But when the number of English feet in an angular day on the equator and the size of the mean Earth degree in English feet are compared, we discover that the two lengths vary according to the Pythagorean comma. That is, in some way Neal's grid makes a connection between the equator's current length and the original mean Earth value.

We can see how the Pythagorean comma arises: The number 360 is employed in the definition of the English foot as 1/360000 of a day on the equator. Neal's grid then establishes the mean length of a degree of latitude as 364953.6 English feet (the length of the degree at fifty-one degrees north latitude). In his system, this degree equals 360,000 "standard geographical English feet," which are therefore longer than the root English foot by the Pythagorean comma.

Based upon these facts, it would then appear that a design criterion for Earth could be to rotate the planet until the variation in length between the ideal mean sphere and the actual length of Earth's equator numerically embodies the Comma of Pythagoras. This seems reasonable, since the Pythagorean comma is the most primordial issue expressed in the study of number and harmony—namely, how to integrate the first two numbers, two and three, within a harmonic set of relations.

THE MYTH OF THE ETERNAL RETURN

Chapter 7 explains that the equator is the only place on Earth that relates space directly to time. There, a unit of measure can be defined that forms the basis for an explanation of how the numerical creation descended into the spinning sphere of Earth. Now that such an objective foot has been defined and measured, we can consider other ancient obsessions, such as the way to return to the state before the creation of sin or imperfection. This state is represented on Earth by the perfect latitudinal degree length (at fifty-one degrees north) of the mean Earth, which is the original temple, the perfect sphere.

This logic runs parallel to astro-mythological concerns in which the spring and autumn equinoxes represent the original state before the planetary world became angled relative to the celestial equator because the equator and the ecliptic, on the celestial sphere, still touch there. This is the concept of the mill that used to grind out sweet but now grinds out salt. This logic is not literal; rather it is indicative and exists within a technical genre of myth connected to measuring degrees of latitude and the length of the equator.

The building of monuments has various harmonic consequences, and the development of a geodetic system of measures makes it possible to scale monuments to suit latitude and perform harmonic calculations using different versions of the same measure, corrected for the Comma of Pythagoras. The same could be said about the demiurge or "builder," since Earth appears to have been designed to incorporate the Pythagorean comma.

The number 3168 is found in the fraction for the metrological comma, 3168/3125, and the numerator 3168 is called "the Perimeter of the Temple" by John Michell because it crops up in both sacred geometries and the perimeters of actual monuments. The denominator fully expresses the number five, as 3125 is five to its fifth power (5^5). As Michell points out in his exemplary *Ancient Metrology,* 31,680 miles becomes the "distance around the sub-lunary world" in his famous "Squaring the Circle" diagram, incorporating the mean Earth and Moon sizes. This "sub-lunary world" is itself a code for everything below the "orbit" of the Moon where, in sacred geometry, the Moon touches Earth, as in figure PS.2. The Stonehenge lintel circle is 316.8

Figure PS.2. The Perimeter of the Cosmic Temple (after John Michell)

feet in circumference and "the ancient [metrological] model was adapted to Christianity" says Michell, by codifying the words Lord Jesus Christ to equal 3,168 in a gematria letter–number equivalence.

These numerical messages hold good prospects for recovering the science behind codified ancient monuments, employing variations of standard measures within them and a common system of canonical numbers throughout. What Michell and Neal have revealed is the secret language of measure that miraculously arose in the ancient world—a discovery at least as important as language itself—that forms part of the evolutionary itinerary of the human race but was nearly lost.

The matrix of creation is entirely compatible with the development of such a system of measures. It would have been numerical sky knowledge that developed into the measuring of Earth and the discovery that numerical absolutes such as the Pythagorean comma are embodied in Earth. Thus we can see the universe, with all its elements of imperfection, setback, and tragedy, as a self-revealing manifestation of pure number designed to make the ascent of intelligent life back to its Source possible.

APPENDIX

ASTROPHYSICAL CONSTANTS

Lunar Year	354.367 days (12 lunations)
Eclipse Year	346.62 days
Lunar Node Period	18.618 years or 6800 days

	SYNODIC	SIDEREAL	
	(days)	(days)	(years)
Mercury	115.88	87.969	0.24085
Venus	583.92	224.701	0.61521
Earth	365.242199	365.256	1.00004
Moon	29.53059	27.32166	
Mars	779.94	686.980	1.88089
Jupiter	398.88	4332.589	11.86223
Saturn	378.09	10759.22	29.4577

GEOPHYSICAL CONSTANTS

Polar Radius	3949.921 miles (circ. = 24,816.8 miles)
Equatorial Radius	3963.221 miles (circ. = 24,903.4 miles)
Meridian Circumference	24,883.2 miles
Moon's Radius	1083 miles (ancient value = 1080 miles)

METROLOGICAL CONSTANTS

Imperial foot	12 inches
Imperial yard	36 inches or 3 feet
Remen	14.58 (14.6) inches
Greek cubit	18.25 inches
Royal cubit	20.618 inches or 1.71818 feet
Megalithic yard	32.618 inches or 2.718 feet
Mile	1760 yards or 5280 feet

GLOSSARY

Aphrodite. The Greek goddess of beauty "born from the sea foam" after the emasculation of the first god, Ouranos, and associated with the planet Venus.

Athena. The Greek goddess of wisdom, who emerged directly from Zeus's head and was the patron of Athens and "sister" of Hephaistos.

Aubrey circle. A circle of fifty-six holes within the outer bank of Stonehenge, discovered by John Aubrey (1626–97).

biosphere. The term for an energetic sphere within which life exists.

bluestones. A type of stone from west Wales used in the later phases of Stonehenge.

chronon. The astronomical unit for the extra time taken for the Sun to rise each day beyond one sidereal revolution.

Chronos. The Greek god of time, associated with the planet Saturn.

Comma of Pythagoras. The mathematical mismatch between seven full octaves of ratio 1:2 and twelve perfect fifths of ratio 2:3 in music.

conjunction. An astronomical term for the meeting of two planets in the sky as seen from Earth.

cosmogenesis. The process whereby the cosmos is created.

cosmology. Any theory of how the universe came into existence.

demiurge. The "Maker" of the planetary system or world in Platonic and Gnostic thought.

eclipse year. An astronomical period for the time taken for a given lunar node (where eclipses may occur) to meet the Sun again.

Enuma Elish. Seven Assyrian tablets recounting the Sumerian creation myth, written in the twelfth century BCE.

equinox. The two times of the year when day and night are equal, occurring around March 20 and September 22.

fifth. The musical interval of seven semitones with a frequency ratio of 2:3, noted as the most harmonious besides that of 1:2, the octave.

fourth. The musical interval of five semitones with a frequency ratio of 3:4, complementary to the fifth.

gematria. An ancient science in which words in Hebrew or Greek are analyzed through the numbers associated with each letter.

geodesy. The geometrical description of Earth's shape.

geomantics. The ancient art and science of siting buildings according to a cosmology.

geometer. A person skilled in geometry, used especially in geodesy.
Golden Mean. A unique ratio (1:1.618...) whose reciprocal and square all have the same irrational fractional part as the number itself—i.e. 0.618....
Golden Year (Golden Mean Year). The practical year times the Golden Mean, equivalent to twenty lunations.
heelstone. The stone over which the midsummer Sun rises when viewed from the center of Stonehenge.
henge. A monument made up of markers in wood or stone usually organized into circular patterns or rows.
Hephaistos. The Greek smith-god who is the son of Hera and known as Vulcan in Roman mythology.
Hera. The Greek goddess who is the wife of Zeus and mother of Hephaistos.
irrational number. An infinite non-repeating decimal number that cannot be represented by a fraction.
Jupiter. The fifth and largest planet in the solar system, associated with the Greek god Zeus.
LOP. Abbreviation for lunar orbital period, the time for one sidereal orbit of the Moon.
lunation. The time taken for one cycle of the Moon's phases as seen from Earth.
Lunation Triangle. A right triangle found within the station stones at Stonehenge that approximates the number of lunations in one solar year.
Mars. The fourth planet from the Sun, associated with Ares, the Greek god of war.
matrix. An environment in which something has its origin; the number relations between an array of elements in a system.
mean Earth. The size of Earth if it were not deformed by rotation once per day.
megalithic yard. A measure used in megalithic monuments with a length one third of an inch less than thirty-three inches.
Metonic cycle. The nineteen-solar-year cycle of the positions of the Sun, the Moon, and the stars, named after the Greek god Meton.
metrology. The science of measures, especially those that have survived through both continual use and use in ancient monuments.
nakshatra. An Indian and Chinese system of dividing the Sun's path into twenty-seven or twenty-eight key star patterns.

node day. The difference of 18.618 days between the eclipse year and the solar year.

noosphere. Translated literally from Greek roots as "mind sphere," the term coined by Teilhard de Chardin for a sphere of cognitive energy associated with the biosphere.

numeracy. The aptitude of individuals to visualize and hence perform numerical transformations, akin to literacy.

octave. The musical interval with a frequency ratio of 1:2, hence a doubling in frequency.

opposition. An astronomical term for when two celestial bodies are on opposite sides of Earth or, more generally, of the Sun.

Osiris. The Egyptian god cut up by his brother, Set, and reconstructed by his wife, Isis, and their avenging son, Horus.

Ouranos. The original Greek god of the sky, who was emasculated by Saturn or Chronos; the complement to the Earth goddess.

ouroboros. The Greek iconography of a serpent eating its own tail, representing recurrent cycles.

overrun. The amount of time by which one time period is longer than another.

planetary matrix. A set of rational numerical relations between planetary time periods.

practical year. The number of whole days in a solar year (365 days).

resonance. The interaction of one planet's orbital period with that of another planet through a shared numerical factor.

royal cubits (long royal cubits). A family of measures based upon the English foot divided by seven; the cubit is 12/7 feet long and the Royal foot is 8/7 feet long.

Sarsen ring. The originally complete circle of carved stone lintels at Stonehenge raised by thirty verticals and employing jointed stonework.

Satan. A fallen angel equated with Saturn and Set who precipitated original sin in the Biblical creation story.

Saturn. The sixth planet and second largest in the solar system, equated with Chronos, the Father of Time.

Set. The Egyptian god who was the brother of Osiris and equated with Saturn.

sidereal. Literally "relative to the stars," a measurement made relative to the stars rather than to the Sun's position.

sidereal day. The time between successive risings of the same star on the horizon.

solar year. The time taken for the Sun to return to the same area of stars.

solstice. When the Sun "stands still" at noon on June 21 or December 22, its path reaching its highest in the sky in summer or its closest to the horizon in winter, causing the longest or shortest day of the year, respectively.

standstill. When a planet ceases to move, but specifically when the Moon stops its journey north or south within the year.

station stones. The four stones placed upon the Aubrey circle at Stonehenge that define a rectangle twelve units by four units.

Stonehenge. The most famous megalithic monument in the world, containing unparalleled numerical sophistication.

synod. The period between two successive meetings of planets in the sky, including the Sun and the Moon.

third. The musical interval of three or four semitones with frequency ratios of 5:6 or 4:5, called minor or major thirds, respectively.

Trigon. The sequence of three conjunctions of Jupiter and Saturn that divide up the zodiac into three parts over sixty years.

tropical day. The twenty-four-hour period in which Earth turns once with respect to the stars plus the time taken to catch up with the Sun's motion along the ecliptic.

Tzolkin. The period of 260 days used by the Mayan calendar, which incorporates both its thirteen- and twenty-day weeks.

Vedas. The oldest of the Indo-European scriptures from the Indian subcontinent; the prominent scriptures of the Hindu tradition.

Venus. The second planet from the Sun, called the evening or morning star as it passes Earth, and associated with the Greek goddess Aphrodite.

Zeus. King of the Greek gods on Mount Olympus and associated with the planet Jupiter.

zodiac. The division of the path of the Sun and planets into twelve parts, each associated with major star constellations.

BIBLIOGRAPHY

Bennett, J. G. *The Dramatic Universe.* 4 vols. Gloucestershire, U.K.: Coombe Springs Press, 1964.

———. *Energies: Material, Vital & Cosmic.* Gloucestershire, U.K.: Coombe Springs Press, 1964.

Blackmore, Susan. *The Meme Machine.* Oxford: Oxford University Press, 1999.

Blake, A. G. E. *The Intelligent Enneagram.* Boston: Shambhala, 1996.

Campion, Nicholas. *The Great Year.* London: Arkana, 1994.

Dawkins, Richard. *The Selfish Gene.* Oxford: Oxford University Press, 1989.

De Santillana, Giorgio, and Hertha von Dechend. *Hamlet's Mill.* Boston: David R. Godine, 1992.

Elkington, David. *In the Name of the Gods.* Dorset, U.K.: Green Man Publishing Limited, 2001.

Graves, Robert. *The Greek Myths.* 2 vols. London: Penguin, 1969.

———. *The White Goddess: A Historical Grammar of Poetic Myth.* London: Faber and Faber, 1997.

Gurdjieff, G. I. *Beelzebub's Tales to his Grandson: An Objectively Impartial Criticism of the Life of Man.* New York: Arkana, 1992.

Heath, Robin. *Stonehenge.* New York: Walker & Co., 2002.

———. *Sun, Moon & Earth.* New York: Walker & Co., 2001.

———. *Sun, Moon & Stonehenge.* New York: Walker & Co., 1998.

MacKie, Euan. *Science and Society in Prehistoric Britain.* London: Paul Elek, 1977.

Martineau, John. *A Little Book of Coincidence.* New York: Walker & Co., 2002.

McClain, Ernest G. *Myth of Invariance.* Boston: Shambhala, 1978.

———. *Pythagorean Plato.* York Beach, Me.: Nicolas-Hays, 1990.

Michell, John. *Ancient Metrology.* Bristol, U.K.: Pentacle Books, 1981.

———. *A New View Over Atlantis.* London: Thames & Hudson, 1986.

Michell, John, and Christine Rhone. *Twelve-Tribe Nations and the Science of Enchanting the Landscape.* London: Thames & Hudson, 1991.

Neal, John. *All Done with Mirrors*. London, Secret Academy, 2000.

Ouspensky, P. D. *In Search of the Miraculous: Fragments of an Unknown Teaching*. London: Routledge & Kegan Paul, 1950.

Pliny the Elder. *Natural History*. Translated by J. E. Healy. London: Penguin, 1991.

Schultz, Joachim. *Movement and Rhythms of the Stars*. Glasgow, U.K.: Rudolf Steiner Press, 1987.

Schwaller de Lubicz, R. A. *The Temple of Man*. Rochester, Vt.: Inner Traditions International, 1998.

Sullivan, William. *The Secret of the Incas*. New York: Crown Publishers, 1996.

Tame, David. *The Secret Power of Music*. Rochester, Vt.: Inner Traditions International, 1984.

Thom, Alexander. *Megalithic Sites in Britain*. Oxford: Clarendon Press, 1967.

Warm, Hartmut. *Die Signatur Der Sphaeren*. Hamburg: Keplerstern Verlag, 2001.

INDEX